T0139048

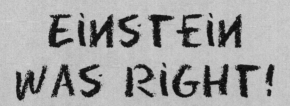

Einstein Was Right!

KARL HESS

PAN STANFORD PUBLISHING

Published by

Pan Stanford Publishing Pte. Ltd.
Penthouse Level, Suntec Tower 3
8 Temasek Boulevard
Singapore 038988

Email: editorial@panstanford.com
Web: www.panstanford.com

British Library Cataloguing-in-Publication Data
A catalogue record for this book is available from the British Library.

Einstein Was Right!

ISBN 978-981-4463-69-0 (Hardcover)
ISBN 978-981-4463-70-6 (eBook)

Printed in the USA

To Walter Philipp and Robert Basler

Contents

Preface

Eritis sicut deus, scientes bonum et malum.

—Vulgate Bible

You will be just like God, knowing the good and the bad.
Also found as the devil's entry in a student's book,
after ridiculing all human knowledge.

—Goethe, *Faust* I

Human fantasy created gods in man's own image . . .
After stating that science without religious attitude is lame.

—Einstein, *Out of My Later Years*

Einstein was godlike for me, as far as physics was concerned. He had this fantastic ability to choose the good assumptions of the physics of his time and to discard the bad ones. He had an unreachable skill. My only hope to make some contribution to physics was to believe in him and to try to stand on his shoulders, much as Heinrich Schliemann had believed Homer's description of Troy's location and then excavated it.

Heinrich Schliemann as a schoolboy used to dream of the famous ancient city that was beleaguered by the Greeks and, so the story goes, conquered by trickery involving the Trojan horse. Schliemann wished to find Troy and the treasures of its leader Priamus. Life took him, however, along a different course and he became a very successful and wealthy businessman. At the height of his success he decided to pursue his boyhood dream and to find and excavate Troy. The professional archeologists first mocked him and denied that the ruins he found were Troy. As we know now, Schliemann did find Troy and he also found a treasure, may be not that of Priamus; a huge treasure nevertheless.

My dream in high school was to become a mathematical physicist and make a major contribution to that field. Einstein was my idol and I read about him whenever I could. I was particularly interested in his writings on quantum mechanics and quantum probability, and I realized that Einstein opposed the probability interpretation of quantum science, the interpretation that the quantum laws of nature do not follow strict causality and instead include chance or probability in their most basic formulation. I saw in Einstein's writings a Trojan horse that he had left for the quantum experts, and I was dreaming that I could release the forces hidden in that horse. It has taken me almost 40 years to come to that boyhood dream. I did have a distinguished career in applied physics and electronics engineering and participated in the foundation of Illinois' Beckman Institute, an interdisciplinary research center. There I heard a lecture by Anthony J. Leggett on Einstein's views of quantum probability and on the opposing views of John S. Bell. It was then that I decided to start working on my boyhood dream in collaboration with my friend, the mathematician Walter Philipp. Now, after more than 15 years of research on that topic, I am convinced that Einstein's Trojan horse was pulled into the castle of quantum theory by John Bell. This book tells the story of how Walter and I tried to release the horse's powers.

Karl Hess

Acknowledgments

Special thanks are due to Hans De Raedt and Kristel Michielsen for their many discussions, comments, and suggestions regarding the physics that is covered, as well as a very careful reading of the manuscript. Ron and Barbara Winters helped with the philosophical content related to probability, the poetry at the beginning of the chapters, and legal questions related to personal stories in books. Richard Humphrey and David Sonne gave me valuable nonexpert advice on how to present the introductory chapters. My son Karl H. Hess helped a lot to improve both the text and the illustrations.

Chapter 1

A Promising Beginning?

Wenn einer der mit Muehe kaum
gekrochen ist auf einen Baum
schon meint, dass er ein Vogel waer,
so irrt sich der.

—Wilhelm Bush

Free translation:

If someone, who just barely,
crawled up the branches of a tree,
supposes then he is a bird,
in awful error is this nerd.

1.1 Introduction

Einstein's Trojan horse, built for the fortress of quantum theory, is contained in a publication that he authored together with two young physicists, Boris Podolsky and Nathan Rosen, and is generally called the EPR paper (Einstein, Podolsky, and Rosen, 1935). Einstein, Podolsky, and Rosen (EPR) presented a proof that the use of probability or chance in quantum theory has very

Einstein Was Right!
Karl Hess
Copyright © 2015 Pan Stanford Publishing Pte. Ltd.
ISBN 978-981-4463-69-0 (Hardcover), 978-981-4463-70-6 (eBook)
www.panstanford.com

unsettling consequences, because quantum theory teaches possible correlations of two distant events. How can single events that are random be correlated to other random single events? EPR suggests that there are only two possible explanations:

- Quantum theory necessitates instantaneous influences, influences that propagate faster than the speed of light in vacuum. These influences of distant events on each other explain, then, the correlations of the two distant events. Einstein was convinced that his relativity theory had excluded the possibility of such influences, and if that was the explanation, then quantum mechanics was simply wrong.
- Quantum theory is incomplete and does not include important elements of physical reality that actually cause the correlations.

Any incompleteness of quantum theory was and still is not acceptable for most of the proponents of this theory that uses chance and probability as one of its basic principles. However, quantum theory also predicts that for certain particle pairs, so-called entangled pairs, the measurement outcome for each single particle is random, while the outcome for the entangled pair may be completely correlated.

You can compare this situation with the case of two "entangled coins." One tossed at one island, say Tenerife, falls randomly on either heads or tails and the second coin tossed at another island, say La Palma, again falls randomly on heads or tails. The very astonishing fact about entangled coins is, however, that both of them fall always on the same side. How can you explain this? One explanation is by some instantaneous influence of one coin on the other over the distance between the islands. The other involves some hidden physical reality such as magnets that make the coins fall randomly but equally on heads or tails at the two islands. A giant struggle of the great physicists of the time followed in the wake of the EPR paper, and Niels Bohr, the leader of the Copenhagen school of quantum theory, disagreed with Einstein and launched massive counterattacks.

In its deepest sense, this struggle of the titans of physics was about the space-time ideas that were used in physics before quantum theory came along. Newton had developed ideas about space and time that included an absolute space, an entity that was just there and could be quantitatively understood and mapped out by certain instruments such as rigid sticks (e.g., a meter measure). Newton's time was something that existed independently of that space and independently of anything else for that matter. Of course, time also could be measured. Newton thought that these measurements could be performed, at least in principle, by some ideal clocks that did not depend on the physical properties of their surroundings. Einstein found significant issues with this space-time picture in his theory of relativity and showed that all clocks slow down when accelerated or subjected to gravitational fields. Clocks that move faster tick slower and stand still when they approach the speed of light. Einstein altogether ruled out motion faster than the speed of light in vacuum.

The attitude of the gurus of quantum theory toward the space-time picture was, particularly in modern times, even more radical than that of Einstein, and some wished to abandon space-time altogether. The Uncertainty Principle, a key principle of quantum theory, works against the precise simultaneous definition of some physical properties of the elementary particles that constitute our world, such as electrons and protons. For this reason, space-time and the strict causality of events lost appeal to some of the leaders of quantum theory and they started to worship at the altar of probability. Einstein never agreed with that movement and thought that without the space-time concept no sensible physical theory could possibly exist.

Years later, with all of the original combatants of this battle passed away, the Irish physicist John S. Bell entered the scene with an astounding notion. Certain statistical correlations of measurements of distant events, precisely those predicted by quantum theory for experiments that Einstein, Podolsky, and Rosen considered in their EPR paper, could not happen if Einstein's ideas of space-time and a limiting speed were correct.

Experimental groups, for example that of Aspect in France and of Zeilinger in Vienna, performed a version of the experiments proposed by EPR. They indeed found statistical correlations that confirm to a large extent the predictions of quantum theory. When Bell saw their results he stated that quantum theory cannot be a local causal theory, meaning a theory like Einstein's relativity. Bell and many of his followers believed that these experiments indicate the existence of instantaneous influences at a distance. For example, when Zeilinger measured pairs of entangled photons each on a different island, one on Tenerife and the other on La Palma, then according to Bell we have an instantaneous influence on a photon in La Palma because of the measurement performed on the photon in Tenerife. For Einstein such influences were just "spooky" and against all the logic that his relativity theory stands for. Also giants of contemporary physics, including Murray Gell-Mann, are opposed to influences at a distance. Gell-Mann used the word "flapdoodle" when talking about science news reports that hinted toward influences at a distance or even teleportation between the islands Tenerife and La Palma. Although I completely agree with Gell-Mann's assessment, I do not think that one can blame journalists for reporting about spooky influences: very serious scientists, including John Bell and many of his followers, do indeed believe that instantaneous influences at a distance do likely exist.

It is my firm belief, and this is the main point of this book, that Einstein's space-time ideas are correct and that John Bell's ideas about statistical correlations are not general enough to be applicable to the EPR type of experiments. This leaves, according to EPR, only one other conclusion: that quantum theory is incomplete. A majority of contemporary physicists do not accept this conclusion of incompleteness. Bell's work has entered all textbooks and is presented as having destroyed any possibility for the existence of Einstein's additional elements of reality.

My opposing views and results presented in this book, developed during collaborations with several excellent scientists, are presented in terms of discussions and personal experiences, not in terms of mathematical theorems. This way, I hope to make the area accessible to everyone who loves science and has a basic understanding of the algebra of numbers. The personal way of presentation and the stories about my family and friends that I have included are to show that science does not exist in a vacuum without human relations or personal problems. There are personality traits involved that may guide or misguide even the greatest scientists, and there are external circumstances that do influence progress. If my friend and coworker Walter Philipp had lived longer, I may have had an even more conclusive presentation. If John Bell would have had a greater aversion to instantaneous influences at a distance, he may not have tried to oppose the EPR work as much as he did in his later years. If my wife would not have supported me and permitted that I was busier with my work than with our move to Hawaii, I would have had to give up my boyhood dream and certainly could not have written this book.

Patience, views, attitude, and even fashion are significant in the history of science. The following example is an illustration of this fact and also attempts to introduce the reader to the basic mathematics that is necessary to understand this book. Mathematics by itself often causes some "attitude," and some will close books immediately when they see mathematical symbols. Publishers of books, therefore, often recommend to authors to use no equations in books that are written for a broader readership, and authors like Arthur Fine, who previously covered the topic of Einstein and probability, did use few or no mathematical equations. I lost this struggle with myself and decided to include a few very basic equations and mathematical symbols. Anyone who understands the following examples will be able to understand the math of the remainder of the book.

My friend Walter Philipp frequently talked to me about the last theorem of Pierre de Fermat that interested him because he worked not only on probability theory but also on the theory of numbers. Fermat's theorem is a great example of how complicated a simple very understandable mathematical question can be to answer. Here is what Fermat's theorem is about.

We know that the following equation is valid:

$$3^2 + 4^2 = 5^2. \tag{1.1}$$

Here we just need to understand the definition of squaring: multiplying the number by itself. We clearly have $9 + 16 = 25$ and the equation above is thus indeed valid. Obviously we can ask ourself the question whether we can find other numbers that we can denote by letters like a, b, and c that fulfill an equation similar to Eq. (1.1), so that we have

$$a^2 + b^2 = c^2. \tag{1.2}$$

The theorem of Pythagoras tells us that there is an infinite number of such equations. A not so trivial question is, however, whether one can find integer numbers (the numbers $0, \pm1, \pm2, \pm3\ldots$) for a, b, and c. The answer is, indeed one can, and one can even find an infinite number of them as the reader can find out from the Internet.

Knowing these facts, Fermat (1637) asked himself whether one could find integer numbers that fulfill the equation

$$a^3 + b^3 = c^3, \tag{1.3}$$

or the more general equation

$$a^n + b^n = c^n, \tag{1.4}$$

for any number $n \geq 3$, meaning a number that is equal or larger than 3, i.e., $n = 3, 4, 5, 6, \ldots$.

This use of $n = 3, 4, 5, 6, \ldots$ complicates the question considerably. Fermat now wanted to know whether one can solve an infinite number of equations and actually thought he had the answer, which was: no, it is not possible to find any integer solution for any equation that uses $n \geq 3$! A surprising answer, particularly because one can find an infinite number of such triples a, b, c for $n = 2$. Fermat published his proof only for $n = 4$, but suggested that there was a proof for all $n \geq 3$. This inspired many people, including laypersons and professional mathematicians, to present a proof and there were at times awards of large amounts of gold offered to those who could. Walter told me that a lot of "proofs" were sent every year to mathematics departments and all were found wrong. The mathematical gurus knew that the proof must be very complicated, and it took until 1994, when Andrew Wiles succeeded with a proof that has been popularized in books and television programs.

Bell's work, or what one calls Bell's inequality, is mathematically even simpler to formulate than the theorem of Fermat because it involves only three mathematical variables, A, B, and C, that may assume only the values $+1$ and -1. Bell's inequality can be written as

$$AB + AC - BC \leq +1. \tag{1.5}$$

Here the symbol ≤ means less or equal. In words, multiplying A with B to obtain the product AB and then adding AC and subtracting BC gives you always a number smaller or equal to 1. You can insert all possible values for A, B, and C and will find that the result is always less or equal than $+1$, exactly as the inequality above states. For example, we can have $A = -1$, $B = -1$, and $C = -1$ and obtain $AB + AC - BC = +1$.

The important fact now is that quantum theory predicts many statistical averages of this kind that do violate Bell's inequality. It is typical for many experiments related to "quanta" or elementary particles such as electrons and photons (the particles of light) that the measurement results give only two possible answers that may be, for example, a click of a detector or no click, or a particle being deflected up or down. These two possible answers are then symbolized by the numbers $+1$ (click or up-deflection) or -1 (no click or down-deflection). Thus, the Bell inequality is about the numbers $+1$ and -1, mathematical abstractions that represent actual experiments. Bell furthermore distinguished three different experimental circumstances (related to the spatial orientation of the measurement equipment) and denoted these three different circumstances by A, B, and C. Because there could be only two outcomes for each circumstance, he assumed that A, B, and C could just be linked to the outcomes of either $+1$ or -1, and thus we have $A = \pm 1$, $B = \pm 1$, and $C = \pm 1$. Bell then took products of the possible experimental outcomes on two different islands such as AB, meaning that you multiply the measured outcome A obtained on the island of Tenerife with the result B obtained on the island of La Palma. Then if we have, for example, $A = +1$ and $B = -1$, the result for the product is $AB = -1$. Bell formed three such products and stated the following. If you perform a large number of measurements and take the statistical average of the products $AB + AC$ and then subtract the average of the product BC, you always obtain a result smaller than or equal to 1. This leads, of course, to the inequality of Eq. (1.5), except that we now talk about statistical averages of experimental results and not just about three numbers like $A = -1$, $B = -1$, and $C = -1$. This fact of statistics being involved complicates any possible proof of Bell's inequality considerably.

How is Bell's inequality related to the question whether Einstein was right or wrong about the use of probability in quantum theory? The Aspect, Zeilinger type of quantum measurements, measurements on two islands that Einstein and his coworkers had suggested as their Trojan horse, result in products AB, AC, and BC and show a statistical violation of Bell's inequality. Bell believed in the absolute correctness and relevance of his inequality and thought that only instantaneous influences at a distance could lead to violations. In a way he was undertaking the giant task of pulling Einstein's Trojan horse into the castle of mathematical physics. The forces that were unleashed by Bell's work led to an enormous number of publications and even to the concept of quantum teleportation. Will these forces tear down some walls of the quantum physics castle and will Bell be shown to be wrong, or will these forces provide a crowning decoration for quantum probability and prove Einstein wrong? This is what this book is about.

1.2 NAS Induction

While walking around Washington DC's monuments and memorials with my wife, Sylvia, I was daydreaming about my work with Philipp. We had researched details of Einstein's writings on quantum theory and probability. Walter was a well-known mathematician working on topics of basic mathematical probability theory, while my past research was mostly about Monte Carlo computer simulations. The name Monte Carlo indicates the major role of chance in this simulation method that is used to explore a wide variety of problems ranging from the stock market to the Brownian motion of molecules. We were particularly interested in the relationship of Einstein's work to the current general excitement surrounding quantum entanglement, quantum probability, and the possibility of building a quantum computer. When we started our work, both Walter and I did not really understand the issues related to quantum entanglement. Now, five years later, we both had learned a lot, and a complete understanding seemed to be just around the next corner. That corner was, however, very far away on this beautiful spring day of 2003, when Sylvia and I turned around the corner of the Lincoln memorial and headed toward the academy building.

The home of the National Academy of Sciences (NAS) of the USA is located close to the State Department in between the White House and the Lincoln Memorial. Sylvia and I had seen the academy building already at the occasion of my induction into the National Academy of Engineering (NAE) in 2001, shortly after the 9/11 attack. Our visit then was very hectic and we had not been able to do much sightseeing, because of the security concerns that had resulted in the erection of concrete barriers blocking most of the government buildings. Now we wanted to see the place in more detail during the day, before this second induction ceremony, this time into the NAS. Fewer than 200 persons are members of both NAS and NAE, and I was very happy and proud of my election, particularly because I was an immigrant from Austria and had only become an American citizen in 1988.

The academy building on Constitution Avenue is unpretentious and in a classical style, to fit to the Lincoln Memorial but not to detract from it. Stairs lead up to a beautiful entrance, to the left of which the Einstein memorial is hidden in trees and greenery. There, comfortably sitting, the hair pillow-combed as known from many photos, a notepad in his hand, Einstein looks relaxed at a star map that forms the statue's basis. The mathematical equations on Einstein's notepad are all about energy and include the so-called Einstein Tensor that relates to the curvature of space-time, the relation between energy E and mass m and the relation between energy and frequency ν. The two equations $E = mc^2$ and $E = h\nu$ are frequently cited, even in the popular press. One cannot think of many equations that have influenced modern physics more profoundly than these, and there are probably not many physicists around who would deny Einstein the distinction of being one of the greatest theoretical physicists of all times. He is the father of relativity theory and one of the fathers of quantum theory.

I was smitten by the equations on Einstein's notepad, and thought how abstruse it was to believe that he had become old and silly and did not understand what great deeds quantum probability and the Uncertainty Principle were. Yet, many of the proponents of modern quantum theory, particularly those working on foundational questions, are very critical of Einstein and maintain that he "blundered" when it came to probability and quantum theory; they think that he was too inflexible to understand the new ways of quantum theory. I detected some satisfaction in the writings of quantum physicists that may have arisen from the fact that they thought they knew better than Einstein. The book by Harald Fritzsch, *You Are Wrong, Mr Einstein!*, with the exclamation mark in red on the cover page, represents one such example. One finds also frequent remarks and quotations of the popular press that Einstein was wrong when it came to probability.

The Uncertainty Principle of Heisenberg is often seen as one of the very few but true stumbling blocks for Einstein's understanding of physics. This principle states that certain "incompatible" physical quantities such as the location in space and the velocity (actually the product of velocity and mass, the so-called momentum) of quantum particles such as electrons, protons, or photons (nowadays also quarks) cannot be determined to arbitrary accuracy in a simultaneous fashion. If the location of an electron in space is known with complete accuracy, then its momentum is totally uncertain and vice versa. Einstein was well aware that any measurement equipment involves, by necessity, quantum particles, often many of them, and may thus be much too "rough" to determine any property of a single particle to arbitrary accuracy. He did not believe, however, that Heisenberg's uncertainty was much more fundamental than the uncertainty of a coin falling on heads or tails after being thrown in the air and bouncing on a rough surface, and he tried to find violations of Heisenberg's "principle." As time went on, however, Einstein realized that the Uncertainty Principle itself was empirically very well proven and theoretically difficult to attack. He still was convinced that probability should not be mixed into the basic laws of nature.

As mentioned, Newton thought that the sun and planets attract each other by instantaneous influences at a distance. However, the work of Faraday, Maxwell, and Einstein had discredited influences at a distance to such an extent that I asked myself why anyone would believe in such influences in modern days. Anthony Leggett, a distinguished colleague at the University of Illinois who later received the Nobel Prize for physics, turned my attention to the

Bell inequality that in his opinion pointed toward the existence of instantaneous influences. After Tony gave a seminar on Bell's inequality, Walter and I started research in this area that led us to believe that Bell was wrong and Einstein was right, and we published that work in *PNAS*, the prestigious Proceedings of the National Academy of Sciences. Our paper caused an allergic reaction of several experts as well as a series of follow-up papers claiming that our work was incorrect. We tried to publish a defense of our work, but *PNAS* refused to publish our defense. Inauguration into the National Academy of Sciences gives, however, the privilege to publish an inaugural paper in *PNAS*, and I decided to submit a slightly improved manuscript as my inaugural paper that was to be distributed before the inauguration ceremony. I was looking forward to discuss our findings with my new fellow academy members and thought about it while sitting under the trees surrounding Einstein's statue.

The clicks of the camera brought me back from my musings about Einstein. Sylvia was taking photos of the Einstein statue. A photo of both of us with Einstein's statue is shown in Fig. (1.1). After this excursion, we dressed for the black tie dinner back in the hotel and then went again to the academy building.

Many people were gathering in front of the entrance of the big lecture room, all busily talking and greeting friends, until an orderly announced that the entrance doors would be opened soon to start the inauguration ceremony. The inaugural papers of the persons to be inducted into the academy were distributed at the door. We were standing for a while in line and received at the door a number of inaugural papers. I looked quickly for my own and finally asked for it. However, my paper was not available. No one knew how that could have happened. However, it was pretty clear to me that this may not have been an innocent accident. *PNAS* had rejected our manuscript, then they had to print it as inaugural. Someone really did not like that.

The main speech before the induction ceremony was supposed to be given by a high official of the Bush administration. Because of the war clouds over Iraq, however, that person could not make it and an academy official gave a lackluster performance instead. I barely listened, because I was still upset that my paper had not been

Figure 1.1 Sylvia and Karl next to Einstein's statue.

distributed. Then the main part of the ceremony followed and the new academy members were called to a podium and introduced and requested to put their signature into the academy book, a fairly thick tome that contains all signatures of all past and present members. Therefore, it also contains the signature of Albert Einstein. I was very proud to put my signature into the same book where Einstein had signed, and I finally forgot my disappointment about the inaugural paper.

Reception, black-tie dinner, and dance were at a different location, one of the beautiful "palaces" of Washington DC laced with so much marble that it compared well to most of the castles of the old world. Sylvia and I walked through a number of rooms, greeting friends and having appetizers and a glass of wine. Actually, I had three or four glasses and was already a bit dizzy when we set down for dinner.

The discussion on the dinner table centered on what science really was and what the role of mathematics was in science. One of the ladies on our table suggested that mathematics was actually itself the greatest science. I had discussed this topic with Sylvia and had convinced her that I knew all about this. Usually it was difficult to convince Sylvia that one was an expert in anything, and her pragmatic mind was capable of uncovering any lack of certainty and detailed knowledge in no time. Mathematics and science, however, were and are not her forte, and she left these areas to me in our well-divided list of capabilities and responsibilities, as I left to her the design of house and garden and the sport activities of our children. She did announce proudly, though: "My husband has thought a lot about these questions," and all eyes around the table focused on me while my eyes saw everyone in a bit of haze. I tried to get away with a short remark and said, "Even Einstein and Mach could not agree on any clear answer. I do not think that anyone really has one." The whole table still looked at me. They expected more, and I therefore gave a short description about my views about science and mathematics.

Ernst Mach defined science as a description of facts of nature that is as complete and as economical as possible. He compared theories to dry leaves and unessential ornamentation. His definition is consistent with some of the great advances of science such as Mendeleev's periodic system of elements. It does, however, not contain a word about mathematics or creativity, not a word about the building of analogies and a precise scientific language, and was in the eyes of Einstein suitable to eradicate "vermin" (mistakes), but was otherwise too stale to be generally useful. In fact, Einstein asked Mach whether he would accept theories such as the hypothesis of atoms if they led to the most economical descriptions of nature, and Mach said that he would. Einstein himself had referred on occasions to Euclid's work as science at its best. Some at the table said that this confirms that mathematics is the greatest science, because Euclid was a mathematician. Einstein, however, had pointed toward Euclid's connection to measurements of objects of the surrounding world. Euclid abstracted the physical surroundings into mathematical idealizations such as lines and dots and angles and then he formulated a few rules or axioms, the smallest possible number of rules in his opinion, to create his geometry. Euclidean geometry was subsequently tested over and over by predictions and measurements and always confirmed. Questions remained on whether his fifth axiom (stating that the sum of the angles of a triangle equals two right angles) was a necessary addition. Einstein found that the fifth axiom was incorrect on a cosmic scale, which followed from his general theory of relativity that was confirmed by measurements of star positions during a solar eclipse. Naturally, Einstein did not see Euclid's work as pure mathematics.

Someone at the dinner table asked me how I would judge modern developments such as the Uncertainty Principle, facts of quantum information, and Einstein's failure to recognize the new and profound truth about it. I started then to talk about my recent work with Philipp and my belief that Einstein was actually right, even in his views about probability, quantum mechanics, and quantum information. However, I was glad when dinner was served and the attention turned to the filet, because I realized that I could not explain my thoughts on these topics during a dinner discussion, not to top experts in science, not to anyone. My views at that time had still too many fuzzy points.

There were two more days of celebrations, visits of Washington DC landmarks, and academy meetings, all very beautiful. Then we returned to Illinois and I went back to work at the Beckman Institute.

1.3 Interdisciplinary Research at the Beckman Institute

The Beckman Institute of the University of Illinois at Urbana-Champaign is an interdisciplinary research institute, and I was one of its founding members. I had served as chair and co-chair of a number of committees who designed a proposal to the Beckman foundation and helped with the architectural design of the building and with its mission, including the choice of the first research groups and faculty. A detailed description was published by Theodore L. Brown (Brown, 2009). Tony Leggett was a member of the founding committees, and I was hoping that he would become involved in the institute's research as one of the leading physicists. The institute and its interdisciplinary mission made me familiar with a wide range of top scientists. Among the initially hired faculty were Paul Lauterbur, who later received the Nobel Prize for medicine, and

Steven Wolfram, the father of MATHEMATICA. The main mission of the institute was to bring scientists from the life sciences, particularly those working on human intelligence, together with scientists and engineers interested in computer-related topics and let them learn from each other and make progress that way. Such interdisciplinary goals are easier set than achieved, and I went on a long learning experience to find this out. John Bardeen, one of the inventors of the transistor and one of my mentors in Illinois, had warned me about it: "If you just sit down and try to perform research on lofty interdisciplinary topics like 'intelligence,' you will achieve nothing. You need specific ideas! We would not have gone anywhere at Bell Labs if we did not have the specific idea to use solids with their much higher density instead of vacuum tubes." I agreed that specific ideas were needed and tried to crystallize several of them.

My successful interdisciplinary collaborations were mostly in areas where engineering and science touch each other and where I had identified some winning ideas mostly related to nanostructures. I did, however, get very excited when Tony Leggett told me that research related to the Bell inequality would be a great topic for the interdisciplinary institute. Tony and I had many interactions before that discussion, and I had hired even some of Tony's very capable postdoctoral researchers. We arranged a seminar that Tony would give on this topic, and this seminar influenced the last seven years of my active career, because I saw a possibility to realize my childhood dream of contributing something in the area of mathematical physics. Tony, I believe now, was more interested in fostering experimental research and to hire someone who would perform experiments. Such experiments would deal with his dreams that were related to Bell's inequality, macroscopic quantum effects, including such exotic topics as macroscopic tunneling, and his fascination with Schrödinger's cat (see the Internet for details). Later, following Tony's initiative, the physics department did succeed to hire Paul Kwiat, a well-known experimenter in the area of quantum optics. Paul is holding a Bardeen chair in Urbana-Champaign as these lines are being written.

The Beckman institute had, from its design, only a minor interest in highly "theoretical" subjects. Beckman started his fortune by constructing a very accurate pH meter that was of great use to the food industry to determine, for example, the pH value of soups. Beckman was the example for greatness in areas where science and engineering touch. Sylvia had a deep intuitive understanding for Beckman's bias toward applications, and Arnold Beckman had taken a liking to her. At one dinner, Sylvia sat next to him and he told her how difficult it was to donate all his money while optimizing the benefit to humankind. Sylvia asked him without thinking, "How much do you have left?" and he told her what everyone at the University wanted to know: "Well, it's mostly in stocks, so I do not know exactly, but it is around four to five hundred million dollars." The ears of the University administrators on the neighboring tables must have gotten a little longer. We did receive a lot of money from the Beckman foundation, 40 million for the building and a lot of additional support for running it. The institute was focusing on human–computer interactions, medical imaging, as well as nanostructures and new types of microscopes that were useful for a variety of areas, even for computer chip technology. None of the themes was as "esoteric" as that proposed by Tony, but I had a very high respect for him as a physicist and wanted to support his ideas.

We had dinner with Tony and his wife, Haruko. Sylvia liked Haruko. Sylvia's feelings about Tony were different: "I cannot say what it is with him, but I do not think that he respects or likes you as much as you respect and like him. Watch out, and don't plan any bigger collaborations and common projects." As usual, I listened to Sylvia, because I had learned in the past that her judgment of personalities was far superior to mine. For me, Tony was a physics star who had my full admiration. I knew, however, that interdisciplinary research takes more than the simultaneous presence of several experts and often fails if these experts do not respect each other and tolerate stupid questions from their colleagues who have a different expertise. I decided to work on the Bell inequality anyway but to start only exploratory research with a part-time appointment for Philipp; no major involvement of Beckman resources. I had worked for many years with Larry Cooper, a highly intelligent member of the Office of Naval Research, and I was sure to get some support from him, because all of this "esoteric" stuff did relate to quantum computing and quantum information. I thought, Cooper would not mind if I used a small amount of money for this type of work, and he indeed did not. The exploratory research that Walter and I performed was on the EPR paper and the related work of Bell and Tony's summary and interpretation of it.

Chapter 2

Einstein's Trojan Horse, John Bell, and Experimental Realization

... timeo danaos et dona ferentes.

—Virgil, Aeneid II

... I fear the Greeks even if they bring gifts.

At the time the EPR paper was written, John Bell was not yet in high school. The EPR paper was aimed against the Copenhagen (Bohr, Heisenberg) interpretation and proposed a crucial experiment that was modified later to make it more feasible. That modified experiment was performed by many groups. Two of these experiments are described below: that of the Aspect group in France and that of Zeilinger's group in Austria. These experimental investigations, their experimental setup, and their measurement procedure were already influenced significantly by John Bell's suggestions and findings. The following sections summarize the EPR–Bohr controversy, the Aspect and Zeilinger experiments, and Tony Leggett's seminar that weaved it all together with John Bell's work. Bell's work and the experimental results form the basis for all claims that the EPR paper

Einstein Was Right!
Karl Hess
Copyright © 2015 Pan Stanford Publishing Pte. Ltd.
ISBN 978-981-4463-69-0 (Hardcover), 978-981-4463-70-6 (eBook)
www.panstanford.com

is flawed and that Einstein was not right with his negative opinion about the use of probabilities in science.

2.1 War with Bohr

There are numerous stories demonstrating how Einstein wished to prove exceptions to the Uncertainty Principle of Heisenberg in his discussions with the great Niels Bohr. The principle is now frequently summarized by the following statement: it is impossible to measure, with absolute precision, two quantities whose product has the dimension of "action." Such quantities are energy and time as well as the spatial location and the momentum of a particle. The constant of nature with the dimension of action is Planck's constant h. The so-called spins of quantum particles also have the dimension of action, and for them the Uncertainty Principle asserts that the spins that one would measure using different directions of spin measurement equipment (magnets or polarizers) cannot be determined simultaneously.

Einstein did, of course, understand that if one had only a fly swatter to locate a fly, then the location of the fly could not be determined to greater accuracy than the spatial area that the fly swatter covered. Many instruments dealing with atoms do have the appearance of a large fly swatter when compared with atoms, electrons, or other basic particles of physics. The existence and interactions of such atomic and subatomic particles are, in the final analysis, always detected through some many-body effects, some interaction and action of a multitude of atoms, and other quantum entities and corresponding fields, which finally leads to some instrument indications that humans can record as data. Einstein was OK with this. It was, however, against Einstein's intuition that there was an *uncertainty as a matter of principle* and that this uncertainty encroached on the basic ideas of space and time. Einstein believed that even quantum particles (such as electrons) did "possess" properties related to both being somewhere in space-time and moving with a certain velocity or momentum. Heisenberg's "uncertainty as a matter of principle" denied a simultaneous reality related to space-location and velocity, and assigned a reality only after a measurement of either one or the other. This idea did not jive with Einstein's views of cause and effect and the fundamental importance of space-time for any type of scientific reasoning. Einstein did not think that any science, and particularly not physics, could stand without the solid basis of space and time and particularly his relativistic space-time that describes the dynamics and ordering of the material world.

A titanic struggle ensued between Einstein, Heisenberg, and Bohr, climaxing at two of the famous Solvay conferences. If one reads the stories about their discussions, one cannot help thinking of little boys who stick out their tongue when they think that they have defeated the opponent. Although it is the picture of Einstein with his tongue out that is known by almost everyone, I think it was actually Bohr who stuck the tongue out on these occasions, at least symbolically speaking.

For example, the story goes that Einstein came up with a detailed model that he believed defeated the Uncertainty Principle. Bohr then showed that Einstein's model was wrong, because it did not account for a result of Einstein's theory of general relativity. How this must have galled Einstein, who knew very well that Heisenberg's theory did, in its design, not include any of the principles of the general theory of relativity. How could Heisenberg's Uncertainty Principle be proven correct by taking recourse to the results of general relativity that Heisenberg never included in the "derivation" of his principle in the first place? Clearly, these discussions were richer than ordinary science-discussions. They contained social aspects and personality conflicts, they contained opinions and fashions of the time, and they contained elements of pride of the scientists for their new developments, and they probably were driven by Bohr's and Heisenberg's fear that Einstein may indeed have found some inconsistency in their work, as he did in some of Bohr's work that was related to energy conservation. When Einstein left these battles and the Solvay conference, the majority consensus of the participants was that his objection to the Uncertainty Principle was defeated. Einstein, however, did not stop thinking about it. He was a great master of probability in physics. He had shown this in his thesis work on Brownian motion and again in his work with Bose and in his early papers related to quantum statistics.

Einstein did learn, from his many attempts to refute it, that Heisenberg's Uncertainty Principle was important, and he expressed this fact later by the following statement: "Heisenberg has convincingly shown, from an empirical point of view, that any decision as to a rigorously deterministic structure of nature is definitely ruled out, because of the atomistic structure of our experimental apparatus." Einstein thus acknowledged (Einstein, 1950) that it is virtually impossible that any future knowledge could compel physics to relinquish the present statistical descriptions. In recent times, it was found that the Uncertainty Principle required very careful scientific definitions that extended both the views that Einstein and the Copenhagen school had at the Solvay conferences (Ozawa, 2003). This modern treatment goes beyond what can be explained in this book. Fortunately, it is not necessary to go into these details. Einstein concocted a Gedanken experiment that addressed the philosophical core of the Uncertainty Principle, but it was in its goal and design reaching beyond that principle. His fight was aimed toward asserting the fundamental role of space-time in physical science and a corresponding deficiency of quantum theory. Years after the Solvay congress of 1927, when already settled in the United States at the Princeton Institute for Advanced Study, Einstein finally got to the crux of the problems that he saw in the formulation of quantum theory and in Heisenberg's interpretation.

Einstein's declaration of war to Bohr and Heisenberg is contained in the EPR paper (Einstein, Podolsky, and Rosen, 1935) and is based on the following idea. There are laws of physics that link velocity and location of two or more particles. A macroscopic example would be the following. Consider two metal spheres (bullets) of equal weight that are accelerated by some explosion in the center of a cylinder (a barrel of a gun open on both sides) that contains them. The spheres are thus propelled out of the cylinder, one to the left and the other to the right, and they fly in opposite directions with equal velocity. Their history is "entangled", because they originate from the same cylinder and are propelled by the same explosion. Then, at some distance to the left one can measure the location of the particle flying left with some location measurement equipment, while to the right one can measure the velocity and momentum of the correlated particle with some different equipment that measures velocity and momentum. Each of the two measurements, of location on one side and of momentum on the other, can be performed without contradicting the Uncertainty Principle because they are not measured simultaneously for the same particle. For a more detailed discussion see (Ozawa, 2003). What Einstein's idea refutes, however, is Bohr's and Heisenberg's interpretation that location and momentum of quantum entities do not have *any* real principal existence, but are only determined in the moment of measurement. The momentum of both particles is known in Einstein's thought experiment, because the momentum is measured on one side. The laws of physics (momentum conservation) give us, then, the value of the momentum of the particle on the other side. In addition, one also measures the location of the particle on the other side and, therefore, has knowledge of both location and momentum. Thus, because of the correlations of the two particles by the laws of physics, we conclude that a particle can and even must "possess" elements of physical reality related to both its position and its momentum.

The EPR paper stated that if one has a physical process whose outcome one can predict with certainty (in probabilistic language with probability 1), then there must be an element of physical reality connected to this outcome. The term "element of reality" referred to the elements of the physical world that Mach gave in his definition of science. Note that Mach's list of elements contains not only objects such as colored marbles but also measurement times (clock times), measured distances and, therefore, space-time. The EPR work thus clearly contradicts the interpretation of quantum theory by Bohr and Heisenberg, who maintained that no such element of reality existed in principle and that the velocity and location of a quantum particle were only determined in the moment of measurement. EPR concluded that quantum theory was thus incomplete, because it did not contain a reference to such elements of reality. The only other way by which such a coincidence of outcomes on two sides could be achieved (not including these elements of reality) was some instantaneous influence over the distance. One could say that as soon as a measurement is made on one side, the other side "knows" instantaneously. For Einstein, this was just a type of spook, and that's what he called it: spooky action. As repeatedly stated, his theory of relativity maintains that there is no higher speed of information communication than that of light in vacuum. Of course, any actual experiment addressing the EPR alternatives of instantaneous influences and/or incompleteness must rule out by its design that the two different measurement equipments (one for position and the other for momentum) can communicate with the speed of light or lower speeds.

Later, in his letters to Schrödinger, Einstein gave very basic explanations of his ideas, and one can imagine that these trivial explanations bothered Bohr. Consider two boxes, only one of them containing a ball. The boxes are closed and one cannot see which of them contains the ball in any way. Einstein maintained that what Bohr was saying was that the likelihood (probability) to find a ball in one of the boxes is 50% (probability or likelihood equals 0.5) and that no other statement could be made before a "measurement" was taken by opening one of the boxes and checking whether the ball was indeed inside. If the measurement (looking in the box) is made and the ball is not in the box, then one knows with certainty that the ball is in the other box. Now, if one asserts that before the measurement the ball is in a "superposition," meaning some state of being in both boxes with a certain probability, and if it does not "materialize" in the box when we open the lid and look, then there must have been some instantaneous "action" at a distance, because the other box some distance away must now contain a ball. Thus Bohr's quantum theory, in its most exacting interpretation, implies instantaneous actions at a distance. If, on the other hand, Bohr would admit that there existed a ball in one of the boxes in the first place and our knowledge changes only after looking, then Bohr's quantum theory was incomplete. For Einstein this simple example highlighted the basic question: Does quantum theory shake the space-time description of physical experience, is it possible to maintain a space-time description, or does the Uncertainty Principle tell us that at the smallest scale space and time become just "foamlike," as some quantum theorists suspect, and no elements of reality corresponding to the location and velocity of a particle exist.

It is important to notice the subtle dynamic design of the EPR experiment that necessitates at least two measurements at locations separated by large distances and also the determination of the measurement times by use of synchronized clocks at the two locations. The EPR experiment was designed to show that one cannot abandon space-time, as some quantum theorists thought, and still discuss meaningful physics. It was also designed to show the pitfalls of action at a distance, particularly when the simultaneity of two measurements plays some role. As is well known from Einstein's relativity theory, what is simultaneous for the observer resting in the reference frame of the measurement equipment is not simultaneous for a moving observer.

Bohr did not believe in action at a distance at all, but was upset about the EPR paper and published a counterpaper. This was mainly a repeat of his previous thoughts on the subject and was, in my opinion and that of many current physicists, not quite hitting the mark. Nevertheless, it was taken as the word of truth by Heisenberg's and Bohr's friends. Einstein was labeled just stubborn and unable to accept the new gospel. The problem for Einstein was that the experiment that EPR proposed was difficult to perform with atomic or subatomic particles, because determinations of location and velocity or momentum of such quantum particles are not easily accomplished. One also cannot shoot atomic or subatomic particles out of a barrel. And then, there was the other problem that was found much later by the Irish physicist John Bell.

2.2 Einstein's Thought Experiment Becomes Reality

2.2.1 *Bohm's Version, Aspect, and Zeilinger Experiments*

The actual EPR type of experiment that was finally performed, long after Einstein's death, was first proposed by Bohm and is therefore denoted in all that follows by EPRB. EPRB involves entangled spins of quantum particles. The word *spin* indicates a connection to rotation, and the spins of photons are measured in units of Planck's constant h divided by 2π, while those of electrons involve division by 4π. Only two values of the spin can be measured, and we denote them throughout by just $+1$ or -1 and drop the multiplication with Planck's constant and the division by 2π or 4π.

The EPRB experiments can be performed with any particles that exhibit a spin property. However, the only detailed experiments with reasonable statistical significance and large enough separation of the two measurement stations have been performed with photons, the particles that constitute light. Many such experiments have been performed by many groups. Here we discuss only the results of two groups: (i) that of Aspect (Aspect et al., 1982), because his group was the first to carefully separate the measurement stations and to rapidly switch the instrument settings (these are important factors as explained below), and (ii) that of Zeilinger (Zeilinger et al., 2007), because they achieved the greatest distances between the two measurement stations in addition to rapid switching of the settings.

The spin properties of photons are linked to the phenomenon of light polarization. We know from Maxwell's electromagnetic theory that light can have the property of being polarized with consequences that everyone who has sunglasses knows. The electromagnetic field of light can oscillate in certain planes, or its direction can rotate (circularly polarized light). If light falls on a water surface, say of the ocean, and if its electromagnetic field oscillates parallel to the surface (in a horizontal plane), then the light is mostly reflected. We can see that effect when the ocean glitters in the sun. Polarization sunglasses do not transmit the horizontally oscillating electric fields of light. If you wear them, the glittering is reduced and you can see the light coming from below the water surface and recognize the white sand or black lava of the ocean floor.

This effect is very well understood as one can see from the fact that we are able to engineer sunglasses. Consider now the possibility that the electromagnetic field of the light does not oscillate in a horizontal plane but in a plane that forms an angle with the water surface (e.g., 45 degrees). Then physicists are used to decompose that light into a horizontal and a vertical component by using vector calculus. The horizontal component is reflected by the water surface as before but the vertical component propagates down into the water below the surface. This decomposition is not as trivial as it may appear to those familiar with vectors. What actually happens microscopically and how the light interacts with the atoms and electrons of the water surface is not simple at all.

The details get even more complicated if one tries to explain polarization effects with quantum theory which regards light as composed of indivisible particles. What does a photon do when it hits the water surface? The quantum explanation contains two basic assumptions. First, the photon has a so-called spin property, or simply a "spin," that relates to the possible polarization parallel and perpendicular to the water surface (or circular polarization). However, instead of relating definitely to either vertical or horizontal polarization, the photon can be in a "superposition" that relates to both polarizations simultaneously and is analogous to the superposition that was discussed above in connection with the ball and the two boxes. This superposition of "quantum states" is a major principle of quantum theory. Quantum theory postulates that such a photon is reflected from the water surface with a certain probability P_1 and transmitted below the water surface with a probability $P_2 = 1 - P_1$ depending on the photon's quantum state or its superposition of quantum states.

Here is how experiments with photon polarizations relate to Einstein's originally suggested experiment with the two bullets flying in opposite directions. One can detect nowadays single photons that are sent out from a source to two different measurement stations. In addition, one can send these photons through polarization glasses toward a photon detector and thus determine their spin properties. As discussed, Bohr maintained that the spin of a particle, including that of a photon, is determined just in the moment of measurement. The crux of the EPRB experiment with photons is that one can create correlated photon pairs, so-called entangled photons. One can even generate a complete correlation of the two polarization measurements at distant locations, and if one side shows a click at a certain time measured by a clock, so will the other measured by a second clock, as long as the polarization glasses of each side have equal setting. The clocks need to be synchronized, and one usually evaluates the measurements within a certain time period, the so-called time window.

How can the polarization property be determined only in the moment of measurement and in a random way, but then be the same for two distant measurements? Einstein claimed that this correlation can only exist if there exist elements of physical reality that are not accounted for by quantum theory. Without these elements of reality, one must invoke instantaneous influences or

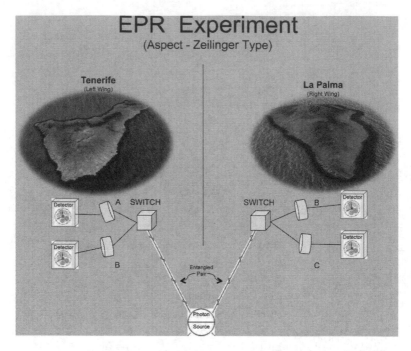

Figure 2.1 EPRB experiment performed by the Zeilinger group on two islands of the Canaries. For more detailed schematics, see Fig. (2.2).

actions at a distance. How else could it be guaranteed that two measurements are completely correlated while the single outcomes are completely random.

In order to address Einstein's objections experimentally, one needs to make sure that the two measurement stations that detect each photon of the entangled pair are apart far enough and the polarizers are switched quickly. Information transfer between the measurement stations that can happen with the speed of light in vacuum or slower needs to be excluded this way, otherwise we cannot apply Einstein's reasoning that involves the separation of the two measurements from each other. A credible separation of the measurements on the two sides was indeed achieved by Aspect and his group in the laboratory, while Zeilinger's group performed even measurements on two different islands.

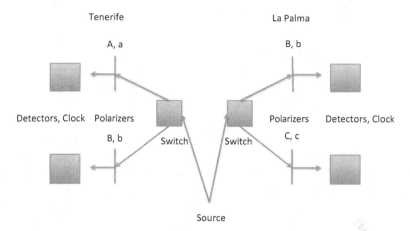

Tenerife   La Palma

A, a   B, b

Detectors, Clock   Polarizers   Polarizers   Detectors, Clock

B, b   Switch   Switch   C, c

Source

Figure 2.2 Schematic of the EPRB experiment as implemented by the Aspect and Zeilinger groups. A source emits entangled photon pairs, which are sent to optical switches located at two distant locations in a left and a right wing of the experiment. The Zeilinger group measured photon polarization on two different Canary islands. In Tenerife and La Palma, entangled photons are received and an optical switch forwards them to polarizers oriented in different directions. For consistency with our previous descriptions, we use the directions **a, b** in Tenerife and **b, c** in La Palma, respectively. If the photons are transmitted by the polarizers, they hit a detector that is also coupled to a clock that indicates and stores the measurement time. Again for purposes of consistency, detection results are denoted by A, B in Tenerife and by B, C in La Palma.

2.2.1.1 Actual experiments by Aspect and Zeilinger

As illustrated in Figs. (2.1) and (2.2), the groups of Aspect (Aspect et al., 1982) and later Zeilinger (Zeilinger et al., 2007) used photon pairs that are emitted from a photon source. This source involves atoms with electrons that are excited to a higher energy state in the Aspect experiment. The electrons cascade from this excited state in two steps, each leading to the emission of one photon. These pairs of photons are "entangled," meaning somehow correlated. As in the original EPR idea, the pairs propagate in opposite directions and they are detected far away from each other. The Zeilinger group used a different method to create the entangled photons and separated the measurements with stations on the two Canary islands Tenerife and La Palma. What is detected on the two islands is the spin property or polarization property of the photons. The polarizers that are used may, for example, filter out horizontally polarized light (denoted by H) and transmit light polarized vertically (denoted by V). Single photons are either reflected or pass these polarizers and then are registered by a "click" of a so-called photo-diode detector. The photon pairs that are used in these experiments can be prepared to be correlated or entangled in different ways. For simplicity in our formulation, we describe mostly one type of these possible correlations, for which both sides indicate either the measurement of the H type or the V type; opposite correlations, one side of H type the other of V type, are also possible.

Aspect and Zeilinger added special features to realize Einstein's "separation" with no information transfer by speeds at or lower than the speed of light in vacuum. Their EPRB experiments have included fast switching of the polarizer settings. Each of the receiving stations on the two islands can direct the respective photon to two different polarizers. Polarizers are characterized by a direction in space. As discussed that may be horizontal or vertical. For the Aspect Zeilinger experiments and their relation to John Bell's work, it is important that the polarizers can also be turned in any other direction. We denote the general polarizer directions by bold letters **a, b, c, d ...** and will use mostly **a**, **b** for Tenerife and **b**, **c** for La Palma. The fact that we use here equal setting **b** on both islands is dictated by Bell's inequality as discussed in the introduction and the next section. If a detector measures a click, e.g., with setting **a**, then a data entry is made as $A = +1$, while for no click we have $A = -1$ and similar entries for the other settings.

How is the EPR separation principle obeyed quantitatively? In other words, how can we be sure that the setting on one island does not influence anything on the other island at the time of measurement? Light propagates in vacuum and also in air with a speed of about 30 cm/ns. Therefore, if the switching of the photons toward the different polarizers is accomplished within nanoseconds in each of the measurement stations of the two wings of the experiment, then these stations need just be separated a few meters to make sure that communication of information between the two wings cannot happen with speeds lower than that of light in vacuum. Such fast switching of the settings can indeed be achieved electronically by use of modern devices. The time of switching and detection is determined by very accurate clocks in the two measurement stations. The Aspect group separated the two wings by about 15 meters for their first experiments in 1982, while the Zeilinger group separated the two wings over many miles between Tenerife and La Palma several years later. Both groups were able to show convincingly that they could detect single photon pairs one after the other while switching the settings fast enough to guarantee the EPR separation principle.

2.2.2 *Proving Einstein Right or Wrong*

One result of these EPRB experiments was that for the equal polarizer setting on both sides, in our case assumed to be **b**, the outcomes are equal about 90% of the times and in some experiments even 99% of the times or more. Because information exchange with the velocity of light and below has been excluded, that correlation can only be explained by influences faster than the speed of light or alternatively by the existence of Einstein's elements of physical reality. Thus, the Aspect and Zeilinger experiments confirm the correlation and the two possible logical choices that EPR gave: influences with a speed higher than that of light or additional unknown elements of physical reality.

Yet, this is not what the groups of Aspect and later Zeilinger claim to have achieved. They claim that they have additional results, lots of data, that proved something entirely different, something related to the work of John Bell and his ideas about the statistics of entangled particles that show that Einstein's suggestion of elements of reality was wrong. As already discussed in the introduction, Bell derived a mathematical inequality about long time averages of measurements with certain settings and claimed that it was impossible to violate his inequality without assuming instantaneous influences at a distance. The measurement results of the Aspect and Zeilinger experiments violate Bell's inequality. This violation was the central theme of Tony Leggett's seminar that is described in the next section. Before presenting the contents of this seminar, I summarize the results of the predictions of quantum theory that do show some remarkable agreement with the experimental results of the Aspect and Zeilinger groups. Remaining experimental uncertainties will be discussed in later sections.

As mentioned, the additional results, that are at the basis of the claim of the Aspect and Zeilinger groups, are of a statistical nature. Quantum theory predicts averages over large numbers of measurements for entangled pairs. Consider polarizer setting **a** in Tenerife and **b** in La Palma and a large number of results $A = \pm 1$ and $B = \pm 1$. One can now take the product AB of a series of these measurements and calculate the average. Averaging means you take a large number N of products, add them, and then divide by N. Quantum theory predicts that the polarizer settings can be chosen such that one can obtain theoretical averages of $AB = \frac{1}{\sqrt{2}}$, $AC = \frac{1}{\sqrt{2}}$ and $BC = \frac{-1}{\sqrt{2}}$, which gives an average of $AB + AC - BC = \frac{3}{\sqrt{2}}$. That is clearly larger than 1. This violates the Bell inequality. Aspect's and Zeilinger's experiments did also yield results that are significantly larger than 1 and thus also violate Bell's inequality.

Anyone who has some experience in science expects now the following sequence of reasoning. We have an experiment with data and we have three theories: quantum theory, Einstein's theory, and Bell's inequality. So we just need to check whether the data agree with the theoretical predictions. If they do, the theories are consistent; if they do not, we have to abandon at least one of the theories. Many brilliant contemporary physicists, including Tony Leggett, N. David Mermin, and the Aspect and Zeilinger groups have opted for sacrificing Einstein, certainly the EPR paper. My conclusion is that it is Bell's work that has to be sacrificed.

Bell's theory claims that such statistical averages cannot be achieved by any conventional way of thinking that does not allow instantaneous influences at a distance. Therefore the Aspect and Zeilinger experiments have shown that influences at a distance are a fact of nature. What I claim is that Bell's theory has serious flaws and it is possible to explain the experimental results of Aspect and Zeilinger in a way consistent with Einstein's relativity and without instantaneous influences at a distance, so that at the end Einstein was right. The way to come to that conclusion is long and arduous. It also is very interesting and teaches a lesson about scientific thinking and logical circles that have plagued science since it has existed.

I am very reassured by the fact that the more recently developed quantum theory of consistent histories (Gell-Mann, Griffiths, Hartle, Omnes, and others) explains the quantum-theoretical result also without any influences at a distance.

There were several sensationalized news media reports of the experiments by Aspect and his group (Aspect et al., 1982), and later by Zeilinger and his group, particularly after their measurements at two different islands of the Canaries (Zeilinger et al., 2007). Journalists generally thought that these experiments prove influences at a distance directly: something is done in Tenerife and has instantaneous consequences in La Palma. The journalists stated that Einstein was therefore proven wrong by these experiments. However, what was actually done in these experiments was the accumulation of many measurements over long periods of time and a deduction from the statistics of these measurements that influences at a distance must have occurred. To come to this conclusion, one needs to invoke the intricate mathematics and physics of John Bell's theory. Einstein has never been proven wrong by a single pair of measurements. Only a combination of many measurements with the hypothesis that violations of Bell's inequality require instantaneous influences at a distance "prove" Einstein wrong. However, it may be Bell's hypotheses that are incorrect or at least not general enough, and the goal of this book is to show that this is indeed the case.

The common saying comes here to mind that with statistics one can prove anything, and it is astounding for me that John Bell was not extremely concerned by the fact that he deduced instantaneous influences from statistical correlations that involve measurements over long periods of time. The only reason that I can see why Bell was so sure of his conclusion is the fact that the measurement outcome for the spin on each side appears completely random. This means that the spin on a given side behaves like a fair coin falling on heads or tails with a 50% chance (for spins one denotes the two possible outcomes by $+1$ and -1). This resonates, of course, with Bohr, who maintained that the spin is only determined in the moment of measurement. The important additional point of Einstein and of the Aspect–Zeilinger experiment is that, for equal polarizer settings on both sides, the outcomes of the measurements are correlated and are with high probability the same. This seems to be the reason for some to believe in influences at a distance. How can the outcomes otherwise be entirely random on a given side and strongly correlated to the other side? Einstein was disgusted by the assumption of influences at a distance and, therefore, insisted that some element of reality must determine the spin outcome and the randomness must arise from some facts that one can explain without influences at a distance. As we will see below, one indeed can (see our function rm).

Walter made some funny remarks about correlations at a distance. He claimed that one of his absentminded colleagues had put on one brown and one white sock. When he looked at them later he exclaimed, "That is strange. At home I have another pair like that." "What a great proof for action at a distance," Walter would say. Jokes like this hint toward the fact that the entangled particles have met previously and that they, therefore, can be correlated without any magic being involved. As we will see, however, Bell was right that the correlations of EPRB experiments are a lot more complicated to explain.

2.3 Leggett's Seminar Presentation

Here is the conundrum that Tony explained to us in the seminar, as he presented a summary of Bell's original work. This particular way of discussing Bell's inequality is also presented in Tony's book (Leggett, 1987). Tony explained the EPRB experiment as we have described it above and then introduced two major assumptions. First, he reformulated Einstein's postulate that there exist elements of physical reality that determine the measurement outcomes by stating that the entangled photons of EPRB experiments "possess" an outcome for each different setting of the polarizer. According to Tony, one therefore obtains for a given entangled pair specific values for A, B (in Tenerife) and for B, C (in La Palma), for example, $A = -1$, $B = +1$, and $C = +1$. Note that at this point Tony introduced the following hypothesis: because any of the settings could have been chosen, an actual set of values such as $A = -1$, $B = +1$, and $C = +1$ would indeed have been obtained for a given photon pair. From the three values $A = -1$, $B = +1$, and $C = +1$ one can calculate the products AB, AC, and BC and obtains

$$AB + AC - BC = -3. \tag{2.1}$$

For different photon pairs, one obtains then, according to Tony's reasoning, different values of A, B, and C and a different result for the products listed in Eq. (2.1). Inserting all possible values for A, B, and C in Eq. (2.1), one finds the results -3, -1, and $+1$, but never $+2$ or $+3$. This fact can be restated in the form of a so-called inequality:

$$AB + AC - BC \leq +1, \tag{2.2}$$

where the symbol $\leq$ means "less or equal." In other words, A multiplied by B plus A multiplied by C minus B multiplied by C is always smaller than or equal to $+1$; it can never be $+2$ or $+3$.

Tony looked at his audience and emphasized that whatever outcome of either $+1$ or -1 was inserted for any of A, B, and C, the inequality was fulfilled and the statistical average over a large number of such inequalities fulfills, therefore, a corresponding identical inequality. Denoting the statistical average of any mathematical expression by aver.(*mathematical expression*) we have

$$\text{aver.}(AB + AC - BC) \leq +1. \tag{2.3}$$

Tony stated further that each of the products can be averaged separately over all entangled pairs to give in essence an inequality of the type that Bell derived and that bears his name:

$$\text{aver.}(AB) + \text{aver.}(AC) - \text{aver.}(BC) \leq +1. \tag{2.4}$$

An astounded murmur went through the room when Tony stated that quantum mechanics predicts a violation of that inequality for the long time average over the three products and that the sum of the left side and its average can exceed 2 for certain settings. He also pointed out that the measurements of Aspect and his group indicated a violation of the inequality that was, statistically speaking, very significant. As mentioned above, the maximum violation that quantum theory gives is $\frac{1}{\sqrt{2}}$ for the setting pairs $(\mathbf{a}, \mathbf{b})$ and $(\mathbf{a}, \mathbf{c})$ as well as $-\frac{1}{\sqrt{2}}$ for $(\mathbf{b}, \mathbf{c})$, resulting in $\frac{3}{\sqrt{2}} > 1$ for Eq. (2.2). The

EPRB experiments of Aspect and Zeilinger did not quite reach that level of violation, but certainly a violation that is high enough to be significant. Therefore, Tony stated, both quantum mechanics and the experimental results are at odds with Einstein's claims. Tony identified Einstein's assumption of elements of reality and limitation to the speed of light as the main reason for the validity of the inequality.

Obviously, if there are instantaneous influences by the two sides on each other, then the outcome A in Tenerife might be different depending on what kind of setting was chosen in La Palma and Bell's inequality was not valid. Then, the A in the two products AB and AC could be different because the setting on the other side is different. Many scientists, including Tony, do indeed favor explanations that all of Bell's assumptions are general and valid. They favor explanations involving quantum nonlocalities and maintain that in the moment a measurement is performed in one wing of the experiment (Tenerife), and the spin of the photon is found to be, for example, $+1$, this fact influences the entangled particle in the other wing (La Palma), and if measured with the same polarizer setting, the result there will then also be $+1$.

I never believed in instantaneous influences at a distance and was entirely convinced that Einstein was right and something was wrong here. Walter turned to me and whispered, "If something can be proven in such a simple way, no violation of that math is possible and no great science can be behind it. The letters A, B, and C correspond to different experiments, and only two polarizer settings can be used for any given pair. Why should the A of the first term be the same as the A of the second, and why should the B's and C's in the different product-terms be identical? The measurements are done for different entangled pairs! The fact that the settings could have been chosen differently for a given pair means nothing. Even in a legal procedure they do not admit what could or should have been. No judge would admit Tony's reasoning in court."

The grin on Walter's face was sardonic, and he had made a lot of noise during his comments. He did not suffer easily what he thought were foolish claims. The students in the lecture hall looked at us, and I was a bit embarrassed and signaled Walter to be quiet. Meanwhile Tony got to his important conclusion that there are no elements of reality, no hidden parameters, and Einstein, therefore, had a problem. He looked unhappy while stating this and said something to the effect that he rather had wished to show that Einstein was right.

Walter had warned me on other occasions about believing in oversimplified mathematical proofs of very general propositions. His very strong feelings came from his detailed knowledge of Fermat's theorem and the many failed attempts to prove it. "Everyone on the street can understand these proofs of Fermat's theorem," Walter used to say, "except for those who are educated in mathematics." Now we know that the proof of Fermat's theorem is indeed possible but enormously complicated. Interestingly enough, expert mathematicians have always laughed at short new proofs of this theorem, while many good physicists hail any new short "proof" that Bell was right and Einstein was wrong.

Bell's followers usually do indicate respect for Einstein and state that his physics is very appealing, but then add that Bell clearly showed that Einstein was wrong. Bell himself may have had respect for Einstein, too. However, there is no revelation of love for Einstein in some of Bell's writings (Bell, 2001). Take, for example, his paper "How to Teach Special Relativity," which does not use any of Einstein's ideas to explain relativity. It starts with an example that is outside the realm of Einstein's special relativity because it contains accelerations and centers on Fitzgerald contractions as well as the approaches of Larmor, Lorentz, and Poincare. With Bell's Irish background, it may not be surprising that Fitzgerald is given the deserved appreciation, but one asks oneself what was "eating" John Bell when teaching relativity without reference to Einstein.

Walter and I left Tony's lecture with the resolve to get to the bottom of this conundrum, and Walter got Bell's original paper (Bell, 1964) from the library and handed me a copy.

Chapter 3

The Devil is in the Detail

Quantum mechanics cannot be embedded in a locally causal theory.

—John S. Bell, La nouvelle cuisine

Subtle is the Lord, but devious he is not.

—Albert Einstein

When I took a first glance at Bell's paper, I got the impression that all of Bell's statements were mathematically straightforward and simply true for numbers and, therefore, a mathematically proven truth, a theorem for numbers, just as Tony's equation was. Both Walter and I could not see why this theorem could be applied to the actual experiments. I realized that it would be a lot of work to understand the physical significance of Bell's ideas, and I first had to convince myself that it was really worthwhile. What I asked myself was whether Bell's ideas had any consequences for engineering and technology and not just for the esoteric questions related to the battle between Einstein and Bohr. While I still was fond of my childhood dream to make a contribution in mathematical physics, I had worked too long in areas of applied physics and engineering to attack a problem that had no applications in the foreseeable future.

Einstein Was Right!
Karl Hess
Copyright © 2015 Pan Stanford Publishing Pte. Ltd.
ISBN 978-981-4463-69-0 (Hardcover), 978-981-4463-70-6 (eBook)
www.panstanford.com

The single most important application of Bell's ideas that I could think of was the use of entangled particle pairs for quantum computers. Therefore I read a few papers on quantum computing. The basic ingredient of quantum computing is the so-called qubit. A photon of the Aspect and Zeilinger experiments is an example of such a qubit, because of the following property. Any measurement of the photon with a polarizer will result either in no click of the detector indicating a "0" or a click indicating a "1." The bit of conventional computers also represents either a "0" or a "1." The qubit, however, has another special property. The actual value that the qubit assumes occurs with a certain probability, and probabilities are characterized by a continuum of real numbers. The probability that someone wins an election may be 2.534%, corresponding to the number 0.02534. Repeated use of the qubit comes, therefore, not only with the digital information but also with some link to real numbers, which are linked to so-called continuous information. This fact, which is always emphasized by the proponents of quantum computing, means that the qubit carries more information than the bit.

This fact by itself did not impress me, because I knew that there existed combinations of so-called digital and analog computing that accomplished the same. Analog means simply a machinery that indicates a continuum of values, such as light of different intensities detected by a camera or some electrical current in a transistor. Such an addition of continuous information is not special. What is special in all the papers of quantum computing is the use of entangled qubits such as the entangled photon pairs. All works on quantum computing claimed that this entanglement cannot be emulated by conventional computing. Most papers on quantum computing cited Bell's work as proof for this special behavior. Without the Bell inequality there was (and is) no reason to regard the entangled qubits as something new, as something that could not be done by the conventional digital and analog machines. This convinced me, for the time being, that it was important to study Bell's paper with great care. The possibility of quantum teleportation did not appear credible to me at that time and I looked into it only later as reported in Section 6.2.

3.1 Bell's Paper

Bell introduced symbols A that may assume values of ± 1, just as Tony did in his seminar. In contrast to Tony, however, he introduced these symbols as functions of two variables. Functions are mathematical tools that map any set of things, so-called variables, onto numbers. You can imagine a function as a machine, or an application of your personal computer, that digests objects that are provided and returns a number, for all of our purposes either $+1$ or -1. Bell denoted one variable by λ. The symbol λ represents Einstein's elements of reality that lead to the outcome of the measurement in question. The set of settings of the instruments that are used to perform measurements (polarizers for photons) form Bell's second variable and are denoted by the boldfaced letters **a**, **b**, **c** and **d**, which we have used already previously. Thus, Tony's random variables A, B, and C are in Bell's notation functions $A(\mathbf{a}, \lambda)$, $A(\mathbf{b}, \lambda)$ and $A(\mathbf{c}, \lambda)$, which can assume the values $+1$ or -1.

Mathematics and physics have developed powerful formalisms for the use of functions. All of these methods, for instance the theory of partial differential equations, which is the basis of much of current physics, assume tacitly or explicitly that the variables of the functions are independent of each other. This means one can pick and chose them independently out of different so-called domains of these variables. Therefore, in our above example, each setting of the domain of settings **a**, **b**, **c**, ... can be used together with any λ of the domain of the elements of reality and one can use any given λ with all settings as well as any given setting with all λs. The use of the same λ for all the setting pairs that occur in Bell's inequality is thus automatically granted as soon as the variables of his functions are assumed or implied to be independent mathematical variables.

Thus, in Bell's notation, the inequality of Eq. (2.2) reads

$$A(\mathbf{a}, \lambda)A(\mathbf{b}, \lambda) + A(\mathbf{a}, \lambda)A(\mathbf{c}, \lambda) - A(\mathbf{b}, \lambda)A(\mathbf{c}, \lambda) \le +1. \quad (3.1)$$

This equation is as easy to understand as Tony's, the only difference being that Bell's functions A are functions of two variables that you can think of as computer apps that turn λ and a setting (e.g., **a**) into either $+1$ or -1.

Bell stated that λ could be virtually anything, even a whole set of mathematical and/or physical objects. He claimed that every λ is independent of the settings **a, b, c,** ... because of the fast switching of the settings just before the entangled pair arrives. Because the entangled pair has left the source much earlier than the switching occurs, there cannot be a connection between the particles properties and the setting, and thus between λ and the settings. Bell and his followers discarded models and arguments of opponents that introduced dependencies of λ on the settings by calling these dependencies a "conspiracy." Indeed, a large number of anti-Bell writings can be logically excluded, because they require that λ and the settings perform some contrived correlated dance. The important point of Bell is that the settings can be and are chosen randomly and the correlated "dance" becomes physically unreasonable only explainable by a crazy conspiracy. As we will see, however, and this is the topic of several sections that follow, the situation is not as straightforward as Bell assumed. One may encounter dynamical effects of the measurement equipment and its interactions with the incoming particles over which the experimenter has no control.

The independence of the λs from the settings is thus questioned in the following, in spite of the fact that this independence appears to be guaranteed by the free will of the experimenter. Failures of assumptions of independence are often encountered in interpreting statistical data, and the dependence of variables is sometimes very hidden. Einstein did know about this fact from his work with Bose, which forms a great example for the discovery of not so obvious dependencies. As is well known, Boltzmann had developed the understanding of the statistics of the atoms or molecules of a gas by assuming that these particles constituting the gas could be treated as independent. Boltzmann's statistics fails, however, to describe the physics of gases in general and did not describe the statistics of photons or all particles that are now called Bosons. Bose and Einstein recognized that the particles cannot be treated as independent. They came up with a new statistics, the Bose–Einstein statistics. This was a very "hidden" dependency that led to a very different statistics from that of Boltzmann. Dependence and independence is, therefore, a topic that requires great care, particularly when it comes to quantum particles. Walter and I had soon a feeling that here was the crux of the problem. But it took a long time to figure out the complex details related to both the mathematics and the physics of Bell's inequality.

I believe it is fair to say that the picture that Bell had for his elements of reality λ was that of colored and flavored marbles with each entangled pair showing correlations in the coloring or the flavor. Remembering, as Bell surely did, that modern high-energy physics treats elementary particles in similar fashion and assigns colors and flavors even to quarks, this picture appears general enough and the independence of the λs from the setting appears justified. As we will see, however, it is not.

3.2 First Suspicions about Bell's λ

As a researcher in the area of Monte Carlo simulations, I had an immediate problem with Bell's general use of λ. If λ could be anything, how did we know that λ and the settings **a, b, c, ...** were mathematically speaking independent variables, which was an absolutely necessary condition for the mathematical formalism that Bell used. "Anything" is not independent of "something." As outlined above, Bell used λ and the settings **a, b, c ...** repeatedly and prominently as independent variables in the mathematical operations that he performed when deriving his inequality. In fact, all of the proofs of Bell and his followers make ample use of the independence of λ and the settings **a, b, c ....** As can be seen from Eq. (3.1), every λ needs to occur with equal frequency with all three setting pairs when many measurements are performed. Otherwise the measurement results cannot be ordered in such a way that one has identical λ for each three setting pairs in Eq. (3.1). Inversely, if one encounters the same λ for each triple of the setting pairs, then one can always reorder the experimental results so that one obtains equal λ, just as shown in Eq. (3.1). Thus the assumption that λ and the settings are independent mathematical variables is already a necessary and sufficient condition for the Bell inequality. But how could anyone prove such an independence if λ can be anything? Walter put it differently: "How does Bell know that all the data can be ordered into inequalities, each having six equal λs? This is just what needs to be shown! How does he know that two different experiments with the same setting can be represented by the same function?"

3.2.1 *Generality of Bell's λ?*

I went to Tony's office and asked about λ. "Hmm," Tony said, "λ can be anything." By this time I started to believe that Walter's contempt was justified and asked further: "Why does it appear in every one of the terms of the inequality identically? Is this not an element of reality that can be different in every experiment?" Tony looked slightly startled! This type of silly question was his reward for talking to ignoramuses who wanted to do interdisciplinary research. "Well, Bell would not have been surprised by that question," he said emphasizing the word "surprised," and he explained that the point was that only one term of the inequality needs actually be measured. The other terms can be inferred, because the correlated pairs "possess" these properties denoted by λ. If the experimenter had turned the polarizers or any measurement equipment to other settings, then one would have obtained the other results with that same element of reality λ. I felt stupid and left to tell Walter. Walter burst into laughter and said, "If a restaurant has an elaborate menu and thus 'possesses' a whole bunch of food, do you have it all in your stomach, because of eating one meal a day?" I felt stupid again; interdisciplinary research was hard. I realized much later that Tony was convinced that if the outcomes of the experiments with different settings could be predicted even without actual measurement, then the long-term experimental results would somehow contain all these outcomes. Walter, on the other hand, knew that mathematically speaking one needed certain mathematical assumptions and theorems to come to such a conclusion.

"Walter," I said, "we have to start seriously working on this! Let's go and have lunch at the China restaurant on Green Street and talk it over." Walter did not look very convinced that he should waste his time any further, but he never declined a meal. I think he finally agreed to do some more serious work just out of friendship. Friendship, trust, and mutual respect are very important components of interdisciplinary research. Funny enough, the fortune cookie that we had at the end of the meal read "You will solve an important problem," and that sealed our deal to work seriously on this research topic. A few nights later we had dinner at our house with Walter and his wife, Ariane. Sylvia was enthusiastic about the project. She liked Walter. Walter liked Sylvia and her roast duck, and this was the start of a roller coaster of exciting research with lasting endurance based on a strong friendship.

3.2.2 *Instrument Parameters*

After studying details of Bell's paper and the Aspect experiments, Walter suggested that we should think of the experimental non-idealities and blame them for the conundrum. Detectors did not always work and the clocks that determined measurement time could be a little off. Some elements of reality may have escaped detection. Walter argued that, therefore, we could take just a selection of the λs and explain the conundrum that way. As we found out later, there is a very general argument by Pearle (Pearle, 1970) that indeed can lead to violations of the inequality based on the discarding of certain data. However, I did not like to use Walter's suggestion anyway. I knew that the experimentalists were taking great care to exclude all "easy" violations of Bell's inequality, and I was convinced that something was wrong with Bell's way of dealing with probability theory. Einstein did not like probability concepts in the basic laws of nature to start with, and what Bell did looked to me as too intuitive and without firm mathematical basis.

What I came up with after a few weeks of thinking was the introduction of equipment parameters λ_a, λ_b, and λ_c. Bell's λ came from a source and could not depend on any of the equipment settings. However, I thought that it was plausible to assume that each of the entangled photons was "shaking up" the electrons and atoms of the polarizers. Thus the elements of physical reality connected with the entangled pair could "mutate" through this interaction with the polarizers into some element of reality that depended on the setting of the polarizer it interacted with. I played around with these equipment parameters and thought of them as something like colored or flavored marbles that adopted some additional properties from the equipment settings. Walter liked the idea and suggested that we send a note to the Austrian Academy of Sciences, of which he was a corresponding member. We later found out, however, that my idea was not new at all and could already be found in the literature, most prominently in the work of the Italian mathematician Luigi Accardi, who had called the mutation of Bell's λ into equipment dependent parameters the "chameleon effect." We soon convinced ourselves that this chameleon effect contained also a difficulty, because one could not guarantee that the chameleons would always lead to equal outcomes for equal settings on each side, which is demanded by quantum theory. The chameleon on one side could not know what color the other side had assumed. So how could one get equal outcomes for equal settings on both sides? Only if the chameleon always followed the local settings and had no dynamics of its own. We withdrew our manuscript just in time before it was printed.

3.3 Time is Special

I did not entirely abandon the idea of equipment parameters for a while, but any playing around with them always ended up in fulfilling Bell's inequality, except if one ignored the requirement that equal settings always lead (for each entangled pair) to equal outcomes. But that requirement could not be violated for any physical reason when playing with colored and flavored marbles and now even with chameleons. Indeed, if one represents the source (of the photons in the real experiment) by an urn that contains colored and flavored marbles of any kind that are sent out and measured by any equipment, one can show that the Bell inequality is always valid, provided one enforces equal outcomes for equal settings.

This brought me back to Tony's claim that λ could be anything and to the question, how a variable could be "anything" and at the same time be mathematically independent of the equipment settings. Tony gave me another tutorial on that topic during one of our lunches and elaborated on our free will, meaning that we are free to choose any setting immediately before the measurement. Then he went on to explain that absent any conspiracy of the crazy kind, the settings and λ were independent and suggested that I also talk to N. David Mermin from Cornell, who was a top expert for all of this. I believed by now to the contrary that the actual "conspiracy" was Bell's choice of identical λs in each term of Bell's inequality. I also knew from my Monte Carlo simulation experience that the free will of arranging an equipment setting did not mean that one was dealing with independent variables in the mathematical sense. Something could go on within the equipment that the experimenter had no control over.

3.3.1 *Time and Correlations*

This whole situation reminded me of a Shakespearean drama with all the characters making decisions with their free will, and often pretty randomly. Then natural law, time, and Shakespeare were weaving a pattern into all that happened. When I thought about that, the role of time popped into my mind, and a few seconds later space-time. I could not help feeling that Einstein, knowing so much about space-time, had set a trap for the quantum-probability fans. The EPR experiment was dealing with correlations, and correlations implied always some "simultaneity," a concept that had played a major role in Einstein's relativity theory.

The correlated pair had to be measured "simultaneously," or at least at highly correlated times, at two different locations. Einstein knew from his theory of relativity that what is simultaneous for the resting observer may not be simultaneous for a moving observer. That presents a logical difficulty to instantaneous influences at a distance, because how can the instantaneity then be assessed and validated for all observers? Furthermore, because of the addition of probability considerations in these experiments, time needed to be treated in a special way. Everyone knows that time is not a random variable, while Bell's λ was thought to be one; thus λ and time or space-time have to be mathematically distinguished. Clearly, such a distinction needs to be considered by everyone who deals with probability and elements of reality. Is time an element of reality? We will return to this question in more detail in Section 7.1 and note here only that, of course, the times of measurement (clock times) and the locations measured with meter rods, for example, are indeed elements of physical reality. Space and time measurement data relate the objects of our observable world to each other and describe their dynamics. Such data are, therefore, different from elements of reality such as colored and flavored marbles that describe an existing "entity" but not necessarily any relations between similar or other entities.

Space-time describes and relates the elementary particles and the settings of macroscopic equipment to each other. These equipment settings can, therefore, not be regarded as independent of space-time. Our free will to determine settings does not imply their independence from space-time. Einstein knew this essential fact and often marveled about Schopenhauer's profound statement that "anyone can do what they want, but they cannot want what they want."

I was now convinced that I had detected a falsehood in Tony's statement that λ could be "anything" and in addition be statistically independent of the polarizer settings. λ could not be identified with a time variable, because the measurement time was different for each term of the inequality, while Bell had the same λ in each term. The measurement times obey an ordering, while Bell's λs exhibit randomness. Bell's way worked only for λs representing colored or flavored marbles, or elementary particles, not for their dynamics and their relations to the measurement equipment and to each other. Equipment settings and measurement times cannot be treated as independent mathematical variables! If a certain setting is chosen at a given time, no other different setting can be chosen at that same time. This is also true if one uses a large number of different EPRB experiments. Each measurement is then performed at a different space-time coordinate, and once a setting is chosen at that space-time coordinate and the measurement performed, no other setting can be chosen at that space-time. The repeated use of λ in a given inequality and the assumption that the λs and settings are independent variables had formed the cornerstone of Bell's proof; that cornerstone was now suspect to me.

Both Walter and I noticed that quantum theory itself was very careful not to use instrument settings and space-time as independent variables. Walter said, "Schrödinger was a great mathematician. He would not have made such a mistake." Indeed Schrödinger, one of the fathers of modern quantum mechanics, only used time as an independent variable for his wave function, while different equipment settings and corresponding measurements were included in form of various differential operators. That settled it for me: one has to be careful with the variables and must not use a λ that can be "anything."

3.3.2 *Space-Time and Bell's Inequality*

Our idea was then to separate in our mathematics any possible space-time dependencies from the setting variables. I was convinced that we needed to include some form of time dependencies to model the "inner workings" of the measurement equipment (polarizers, magnets) as well as for the particle source. I suggested to Walter that, at first, we should treat each measurement station just like a personal computer that could run a program, an application or app. The EPRB experiment can then be modeled like this. A common source sends out information, for example by sending a laser beam to each of two computers, one, say, located in Tenerife and the other in La Palma. That laser-information represents Bell's λ. Because it can be different for each time a measurement is performed, we attach a number $i = 1, 2, 3, \ldots$ to each of the λs and write λ_i. For simplicity we assume that the entangled pair carries the same information λ_i to both sides.

The EPRB experiment and Bell's functions can then be simulated as follows. The computer in Tenerife has two apps, one with "setting" **a** and the other with **b**. The computer in La Palma has also two apps, one with "setting" **b** and the other with **c**. These apps start running as soon as a signal is sent to the computers. In the real EPRB experiment, this signal would be the detection of the entangled pair. For the simulation, any trigger-signal that supplies the appropriate information related to λ_i will do. Both computers have internal clocks that work precisely the same way and have been synchronized beforehand. During a certain predetermined time interval of short duration, in the real experiments between 1 nanosecond and 100 nanoseconds, each of the two computers randomly picks one of the two settings that are available. If during this time interval, the so-called coincidence window, a particle or signal is detected on both sides, it is assumed that an entangled pair has been detected. The actual detection occurs at some time t_i in Tenerife and at time t_i' in La Palma, both times being within the given time window. Assume that we have the app with setting **a** in Tenerife and the incoming information λ_6, where the number 6 indicates that we perform the 6th experiment. That app then takes the information λ_6 and calculates a result, say $+1$. If the setting in La Palma was **b** during that coincidence time window and the entangled information λ_6 arrived there also, then the app in La Palma calculates also a value, say -1, and we have a product of -1. This way, we have simulated Bell's functions and, in particular, Bell's product $A(\mathbf{a}, \lambda_6)A(\mathbf{b}, \lambda_6) = -1$. If one performs many experiments with randomly chosen apps and a countable number of arbitrary λ_i, one always fulfills Bell's inequality. This proven fact shows that Bell's work is pretty general.

Walter and I extended now the simulation to include dependencies on the measurement time. What we were doing different, and more general than Bell, was to include a time dependence of measurement equipment and source. This means we need to use different apps for different measurement times even if the settings are equal. The app in Tenerife now depends on the settings as well as t_i and is different for each different i. The same is true for the apps of the computer in La Palma. The functions or apps that we now deal with contain three variables and, for the above example, we have $A(\mathbf{a}, \lambda_6, t_6)A(\mathbf{b}, \lambda_6, t_6')$. Actually we cannot write the functions anymore the way that Bell did it, because Bell's notation implies that we deal with functions of independent mathematical variables. Because the settings and times are not independent variables, and because time is not a random variable, we write settings as a subscript and times as superscript. Then for setting **a** in Tenerife we have the function $A_{\mathbf{a}}^{t_i}(\lambda_i)$, and for setting **b** in La Palma we have $A_{\mathbf{b}}^{t_i'}(\lambda_i)$. This is all illustrated in the two boxes below for measurement number i. Each box represents an app, one in Tenerife and the other in La Palma.

```
APP: EPRB Simulation, TENERIFE


          INPUT:

          a, λᵢ, tᵢ

          OUTPUT:

          A_a^{tᵢ}(λᵢ) = +1
```

```
APP: EPRB Simulation, La Palma


          INPUT:

          b, λᵢ, tᵢ′

          OUTPUT:

          A_b^{tᵢ′}(λᵢ) = +1
```

There is actually one more point of generalization that is necessary. The times have to be replaced by space-time labels st_i and st_i' because the measurements are done at different locations. This results in the apps or functions $A_a^{st_i}(\lambda_i)$ and $A_b^{st_i'}(\lambda_i)$. This generalization is only necessary if different EPRB experiments at different places are considered all at once, or if one includes observers that move with different and very high velocities.

Our notation for the functions may appear to some readers complicated and containing too many labels, "too many notes," as the emperor purportedly said to Mozart. However, Bell did, in the opinion of this author, not use enough of them. The experimental EPRB machinery is elaborate, contains many particles (electrons, protons, neutrons, atoms), and may need the additional space-time labels for their dynamic description. I, therefore, ask the readers to be as patient with our function labels as they might be with the many different names in a Tolstoy novel. The labels of mathematical symbols are as important as the names are in the novels, because they determine the story line. Fortunately, we do not always need all the labels. If all the equipment stands still in a laboratory, as it usually does, we may replace space-time just by the number i of the actual experiment. Then, time is just regarded as an order parameter, which provides order as we do when we are counting. It is important, though, to remember that the λ_i may still be random, just the i cannot be. If we replace space-time coordinates st_i by the number i, we obtain functions such as $A_{\mathbf{a}}^{i}(\lambda_i)$ in Tenerife and $A_{\mathbf{a}}^{i'}(\lambda_i)$ in La Palma.

The relabeling of Eq. (3.1) has big mathematical consequences, because now every term of Bell's inequality may be different, at least in principle. If one thinks of the measurement results in terms of coin tosses, then the "coins" become different for different settings, for different λ_i's, and also for different space-time coordinates st_i (or just i). In other words, the machinery that governs the coin-toss outcomes includes both a space-time and setting dependence, but avoids the mistaken assumption of independence of all mathematical variables. Because all terms of the inequality can now be different, we can achieve a sum total of the products as large as 3 and the inequality becomes

$$A_{\mathbf{a}}^{i}(\lambda_i)A_{\mathbf{b}}^{i'}(\lambda_i) + A_{\mathbf{a}}^{i+1}(\lambda_{i+1})A_{\mathbf{c}}^{i'+1}(\lambda_{i+1})$$
$$- A_{\mathbf{b}}^{i+2}(\lambda_{i+2})A_{\mathbf{c}}^{i'+2}(\lambda_{i+2}) \leq +3. \qquad (3.2)$$

This inequality, in contrast to the Bell inequality, is not putting any mathematical restrictions on the outcomes. It is always true. I apologize to the reader again for the more complicated appearance of the inequality. It still involves only simple multiplications of symbols which have a value of $+1$ or -1. Mathematical symbols that model physical effects need to be labeled in such a way that all the physical differences that are important for the outcomes are covered, and this leads to the cumbersome notation.

For the purpose of further simplification, we can return to Tony's notation, but we still need to retain the index i in order to permit different outcomes for different experiments with the same setting. Then we arrive at

$$A^{i}B^{i} + A^{i+1}C^{i+1} - B^{i+2}C^{i+2} \leq +3. \qquad (3.3)$$

Equation (3.3) is different from the inequality discussed by Tony and can always be fulfilled. One of Tony's criticisms of this equation was that this would mean that the outcomes could be different for different sequences of experiments. Actually, according to quantum theory, they indeed may be, and experimentally indeed are found to be different also. Different sequences of measurements may have different outcomes for the same setting. Only the long-term averages stay the same. Walter assured me that keeping long-term averages constant was mathematically no problem for the new equation, and I convinced myself also that there was no physics problem involved. However, Tony indicated also in subsequent discussions, at least as far as I understood him, that he did not like our reasoning. Of course, a lot remains to be shown to indeed prove that our invocation of possible space-time dependencies of the equipment solves all the puzzles surrounding Bell's work.

Chapter 4

Developing a Space-Time-Dependent Model

It all looks random, bit by bit,
And time then weaves a pattern into it.

There were several problems or major hurdles to be overcome to develop a valid space-time-dependent model. The outcomes always needed to be the same for equal settings on both sides, the violations of Bell's inequality that can be obtained need to agree with quantum theory, and then there had to be randomness of the single outcomes on each side.

4.1 Randomness of Outcomes

"The thing that really worries me," I said to Walter, "is the randomness of the single outcomes. The likelihood to obtain $+1$ or -1 is, according to quantum theory and the EPRB experiments, equal on each side. How can one obtain this randomness and at the same time the required correlations of the two sides? This randomness is the reason why Bell was so convinced that influences

Einstein Was Right!
Karl Hess
Copyright © 2015 Pan Stanford Publishing Pte. Ltd.
ISBN 978-981-4463-69-0 (Hardcover), 978-981-4463-70-6 (eBook)
www.panstanford.com

at a distance are involved!" After discussing this for a while, Walter and I came up with the idea that the randomness of the single outcomes may just be contained in some additional physical effect that is overlaid on top of a nonrandom machinery. Using again the two-computer model for the EPRB experiments, we can include an additional app on each computer that flips the sign of the outcomes randomly but in unison on both computers. Such a random flipping in unison is possible, because both computers can have the same internal clock time. That clock time can be set by the laser signal that comes from a common source and represents the entangled pair.

Mathematically speaking, the flipping is accomplished by multiplying the outcomes on both sides by a function of time rm that equals -1 at random times at both sides and $+1$ otherwise. Such a multiplication does not change the products that appear in Bell's inequality because it acts on both sides. Walter asked whether there was any physical effect that could cause such a sign change. I did not know an answer to his question and said, "If we only wish to show that one can do without instantaneous influences at a distance, all we have to show is that two computers can accomplish the task. If two computers can do it, nature's laws can do it."

Walter then developed a rather complete mathematical model. For the expert I add here a few details. Walter introduced very general random functions, $rm(t_i) = \pm 1$ on one side and $rm'(t_i')$ on the other. As noted earlier, t_i and t_i' are the times of measurement on the two different islands, and by $rm(t_i)$ and $rm'(t_i')$ we denote two random functions that assume either the value $+1$ or -1. We choose the functions rm and rm' in such a way that $rm(t_i) = rm'(t_i')$. Therefore the product of rm and rm' equals $+1$:

$$rm(st_i) \cdot rm'(st_i') = +1, \tag{4.1}$$

and all the products of the results on the two islands are unchanged, while the outcomes on each side are randomly equal to ± 1. It is relatively easy to find software that accomplishes this on two computers; for example, MATHEMATICA can do it.

Thus, introducing random functions such as rm and rm', one can consider functions A that are not random at all. One can take care of the randomness for the simulation of outcomes afterwards, by multiplying the outcomes in both wings with the random functions. The functions (or apps) A themselves may be completely deterministic if we so desire. The products $A \, rm(st_i)$ of the functions A with the random function rm will be guaranteed random. This fact represented the first step for our refutation of Bell's proof. There may be some who are concerned about a distinction between ideal randomness as opposed to randomness that can be produced by computers. For Walter and me this was never a real issue.

4.2 Constructing Apps to Give the Quantum Result

Our next goal was to construct time dependencies for the computer apps or functions A that lead to a violation of the Bell inequality. An example of outcomes for the products of Eq. (3.1) that violate Bell's inequality is $+1$ for the terms with the plus sign and -1 for the last term with the minus sign. Because $-(-1) = +1$, the sum of the three terms is then $+3$, which indeed violates the Bell inequality Eq. (3.1).

From this example, we conclude that it must be possible to design computer applications that return values for functions $A_{\mathbf{a}}^{st_i}(\lambda_i)$ and $A_{\mathbf{b}}^{st_i'}(\lambda_i)$, etc., in such a way that one obtains any desired long-term average for the products and, therefore, also the average predicted by quantum theory. Figure (4.1) illustrates possible time-dependent outcome patterns for the return of $+1$ or -1 of such computer applications during some time for a given λ_i and for the given setting pair $\mathbf{a}$, $\mathbf{c}$. The space coordinate of the measurement equipment is assumed to be always the same (e.g., one in Tenerife, the other in La Palma); only the indexes $i, i + 1$ and $i + 2$ are shown to indicate both different λ and times. One can, of course, easily include different apps for different and even moving space coordinates. The outcomes that the computers return are positive and negative over certain segments of (short) time periods, and one can see from the figure that the products $A_{\mathbf{a}}^{st_i}(\lambda_i) A_{\mathbf{c}}^{st_i'}(\lambda_i)$ may indeed be positive or negative according to the design of the computer programs. At the earliest times (lower portion of the figure) and for some given λ_i, the outcomes are all negative no matter at which precise time the laser beams (entangled pairs) hit the computers and start the applications to run. I have included sporadic minus signs $-$ in the

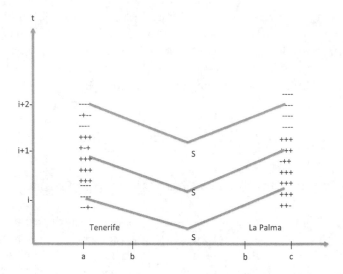

Figure 4.1 Outcome patterns, for example, of functions $A_{\mathbf{a}}^{st_i}(\lambda)$ and $A_{\mathbf{c}}^{st_i'}(\lambda)$ simulated by suitable apps on two separate computers for a given time period. The vertical axis represents space-time st or just time t for the indicated indexes i, $i+1$, and $i+2$. The horizontal axis represents the settings. The pair of lines starting from the source S indicate the entangled photon pair. The times and settings during which the functions are positive for a given λ are marked by $++$ and the times during which they are negative by $--$. The product of the two functions is positive at times when both marked areas are either positive or negative.

sea of plusses $+$ (and vice versa) at certain times, just to indicate that we can include experimental noise and fluctuations in our model. In the second time range (center part of illustration), and for λ_{i+1}, the outcome for the product of measurements can be both positive and negative, and the average over many measurements during this time segment can assume any value between -1 and $+1$. Finally, for the top time segment, the outcome for the product is almost always $+1$. We thus can see that we can write applications that return any value of ± 1 for any given setting pair. In this way one can also violate the Bell inequality for three chosen measurement times, as shown in Fig. (4.2).

The main point of the above examples is that we can find any possible correlations of the measurements for a given time window and these correlations may depend on the settings of

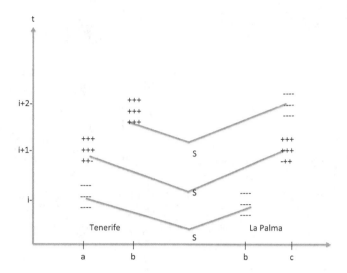

Figure 4.2 Same as Fig. (4.1) now for the three setting pairs used in Bell's inequality. The returned value for the function products is $+1$ for setting pairs **a**, **b** as well as **a**, **c** and -1 for the setting pair **b**, **c**. This means that the corresponding Bell inequality of Eq. (3.1) is violated. Different settings require different photon histories as shown by the exaggeratedly different length of lines from the source S.

both sides without any necessity of instantaneous influences. The reason is that, in addition to the correlations that are determined by the information that the entangled pairs carry, the measurement equipment may add time-related correlations. Clocks in Tenerife and La Palma are not statistically independent. Any time-dependent effects in the machinery of the two stations (and the source) will lead to additional correlations. These additional correlations may depend on the settings of both sides, because of the time window that produces a "mixture" of the tables of both sides. Thus, the time-coincidence window can give rise to setting-, time-, and λ-dependent correlations without any spooky influences.

There was one more problem that we needed to take care of to finish our time-dependent EPRB model. Quantum theory requires that the measurement outcomes are equal for equal settings. The Aspect and Zeilinger experiments, and those of others, do also show approximately equal outcomes for equal settings, although the experimental error can be quite large. Walter solved this problem in a very complicated mathematical way that was criticized by a number of scientists (see below). I believe that a completely satisfactory resolution of the problem was only found much later during my collaboration with Hans De Raedt and Kristel Michielsen, as described in Section 13.1. Walter, however, was quite proud of his achievement and asked me whether I thought that anyone could still object to our reasoning and on what basis. If the Bell inequality can be violated as easily as shown in Fig. (4.2), how can one still believe in it?

I looked at him and said, "I think it is because it cannot be violated for any single given time." One can easily convince oneself that this is true, and Fig. (4.3) illustrates this point. For any given time in the illustration, the products are such that the Bell inequality is valid,

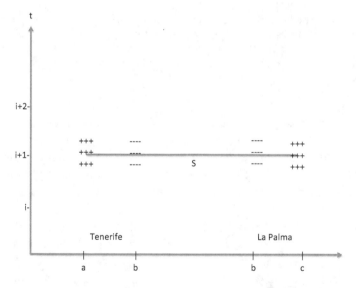

Figure 4.3 Outcome patterns for functions A and settings **a**, **b** on one side and **b**, **c** on the other, assuming that all measurements are performed at the time corresponding to measurement $i + 1$. The returned value for the three function products (results of the apps) always fulfills Bell's inequality, because now the outcomes with both setting **b** need to have the same sign (see text). Note, however, that for any real experiment, only one pair of settings can be chosen at any given time.

provided that the outcomes with the setting **b** on both sides are equal. The reader is encouraged to change the signs arbitrarily with equal signs for setting **b** and see that the Bell inequality stays correct. I was convinced that Bell and his followers just imagined a limit of infinitely dense measurements with all three setting pairs chosen at one particular time. I also was convinced that this limit, which also corresponded to equal λs for each of the three setting pairs, could not possibly be reached. It takes time to switch from one setting to the other. Tony had, however, another more serious objection to our work that is discussed below. Tony called it the Bell game.

4.3 Time-Dependent Physical Processes

Having done the mathematics to his own satisfaction, Walter asked, "What kind of actual physical processes that can cause a time dependence do you have in mind? What is really happening with the photons and atoms in these EPRB experiments?" "I still do not know any physical process for the randomness," I said, "but for the general time dependencies my suspects are many-body processes in the measurement equipment. Think of a polarizer as a large number of atoms and electrons oscillating because of thermal energy and the electromagnetic fields of the surroundings. A photon that falls on the polarizer shakes the electrons and atoms. This shaking causes some random effects like those of an explosion of gunpowder and also some oscillations such as so-called plasma oscillations. The explosion effects are probably random, while the oscillations are periodic processes exhibiting a time dependence and thus can be likened to a clock." "Why would these effects be different for different measurement times?" Walter asked. "Well," I said, "because of the connections with the environment. The earth rotates, and that means something for spins or any type of gyroscope, or because the earth's magnetic field fluctuates, or because the neighboring radio station emits some electromagnetic waves." "No, I mean seriously," Walter said. "Even your famous mathematician colleague John von Neumann suggested that the process of measurement involved some dynamics," I said. "That was a long time ago," said Walter.

"Well, physicists still talk about a 'de-phasing dynamics,'" I said. When Walter still looked incredulous, I tried it with bluster. "I also know that modern physics attributes a lot to the vacuum and there is the idea of local gauge fields that influence quantum particles." Gauge field theories are in agreement with Einstein's space-time ideas. They attach so-called fibers and a phase factor from quantum theory to any space-time point. The gauge field and the quantum phase perform acrobatics on each fiber without any influences at a distance. There may be some connection to what our functions $rm(t)$ and $rm'(t')$ from Section 4.1 accomplish." I did not really know what I was talking about, but hoped that Walter would be satisfied. "Acrobatics?," Walter asked and rolled his eyes. "Is this what I have to deal with?"

So I tried to get down to earth again, away from gauge fields, and explained that the measurements are performed by counting the coincidence of the entangled pairs using clocks and a time window created by these clocks. The timing of the coincidence count may depend on the actual settings. The response of the polarizer electrons and their oscillations may be different for different settings. "Then it is possible", I said, "that a coincidence of the measurement of a pair becomes a function of both settings and the particular measurement times on each side. The interaction time of equipment and quantum particles may, for example, depend on the settings and can cause such a dependence. We can think of details later."

Walter was still not happy with my explanations. "Think of the alternative, Walter," I said. "Think of instantaneous action at a distance." "Tell me more about that," Walter said, and I started to explain to him what Bell's alternative explanation was. Bell believed that it was not possible to explain EPR experiments without invoking some deviation from Einstein's principle that the measurements in the two wings of the experiment (two islands) are without influences from the other side. The reason for Bell's conviction was his inequality and the fact that the EPR experiments of Aspect and coworkers violated it. Bell did not think that his inequality could be violated without some deviations from Einstein's principles. Exactly like Tony, he thought that λ could be anything and his inequality could be written with the same λ in each term. Then one can explain violations of the inequality only by assuming that in the instant of measurement in Tenerife, say with setting **b**, some influence is exerted on the entangled particle in La Palma that changes the given element of reality λ, so that it "knows" that the measurement was performed with setting **b** in Tenerife. Therefore the λs of the inequality are not the same anymore but differ because they include some information about the setting of the other side, if a measurement was done on that side. Walter looked cockeyed at me and said, "So they are accomplishing different λs with spooky influences? How can they disregard Einstein's principle about the velocity of light so blissfully?" "That's a good question," I said. "Einstein's principle is also contained in Maxwell's equations and is proven by all everyday experiences we know."

Remember seeing lightening in the sky and hearing thunder later? This is because the speed of light is so much faster than the speed of sound. We never have any information available that something has happened in our field of vision and then we actually see it happening later. Never ever! This is because the speed of light is the highest speed that we know of. If it were not, we could have information about a murder, about somebody shooting someone, and then later see the actual murder happening. No one knows anything like that! Einstein ruled out such instantaneous information transfer over a distance, because he found destructive evidence against the concept of things to happen simultaneously in general. Events that appear to be simultaneous to some observers that we consider to be at rest are not simultaneous for moving observers, all because of the finite velocity of light in vacuum.

Action at a distance means that if one performs a measurement on one island, instantaneously something happens on another. However, a moving observer would not see these two events as simultaneous or instantaneous. Therefore, in Einstein's opinion, action at a distance could not be a true law of nature (Einstein, 1950). "So how does Bell get around this?," Walter asked. "I don't know whether he got around this," I said, "but what his followers generally claim is that the single measurement outcomes of EPRB experiments are random and therefore no information transfer takes place. It's random on one island and random on the other. So they speak of instantaneous influences at a distance, not of action at a distance." Sounds like the old trick that shows that a turtle is faster than the fastest runner on earth," Walter said.

"I take it that you believe, then, in my explanation of time dependencies," I said. Walter just rolled his eyes again and asked, "Entanglement is, then, in your view just the average of a mess of equipment time-dependencies intermingled with time dependencies of a source that emits correlated quantum particles, while for Bell it is a direct rigid connection of the two quantum particles over large distances. Do you see it that way, and why?" "Yes," I said, "this is how I see it, and my basic reason is that I am totally convinced by Einstein's assessment that there are no instantaneous influences at a distance at all. You have to have faith in some things even if you do science."

4.4 Walter's Status Summation

Walter looked toward heaven, and then he summarized the status of our research as follows. Bell and followers use function products

$$A(\mathbf{a}, \lambda)A(\mathbf{b}, \lambda). \tag{4.2}$$

The functions are defined on a domain of two independent variables, the settings $\mathbf{a}, \mathbf{b}, \mathbf{c} \ldots$ and λs, and can therefore not violate Bell's inequality. Violations of this inequality are blamed by Bell on instantaneous influences at a distance (also called nonlocalities) that cause the λs to depend on the settings of the other side, so that one has

$$A(\mathbf{a}, \lambda(\mathbf{b}))A(\mathbf{b}, \lambda(\mathbf{a})). \tag{4.3}$$

Influences at a distance imply, according to Bell, the functional dependence of the λs on the settings that leads to the violation of his inequality. Entanglement is an effect that exists for every particle pair independent of the measurement equipment.

We, on the other hand, maintain that we may have space-time dependencies of both the correlated particles symbolized by λ and the measurement equipment. The particles that constitute the equipment are "shaken" by the incoming particles and perform a "dance." That space-time dependence leads to the space-time dependence of the functions A to give products such as

$$A_{\mathbf{a}}^{st_i}(\lambda_i) A_{\mathbf{b}}^{st_i'}(\lambda_i). \tag{4.4}$$

This dependence on space-time can lead to a violation of Bell's inequality because all the pair products of functions are now different. "Entanglement" arises in our case from both the correlated particle pairs and the time dependencies of the equipment. It is not an instantaneous effect but the statistical mixing of source and instrument time dependencies over long time periods.

"Hallelujah!" I said. "All we have to show now is that such space-time dependencies have also a reasonable physical basis and describe all the possible experiments." I do not think that I realized what a tall order that still was.

Chapter 5

First Publications

Quod scripsi scripsi.
                              —Pontius Pilatus
What I wrote, I wrote.

5.1 Publication in a Prestigious Journal

I was very enthusiastic about our findings and wanted to publish
them in a prestigious journal. I knew that this was not going to
be easy, because neither Walter nor I were known experts in this
particular area of mathematical physics, and any referee who saw
that we tried to refute Bell would probably have a negative reaction
to it. This is why I called my friend Federico Capasso and asked him
to help. I had known Federico since many years. He was one of the
top researchers at Bell Laboratories and now professor of applied
physics at Harvard. Federico and I had had many spirited discussions
about research topics and had become very good friends. Federico's
research area is solid state physics. He is a great expert in the area
of light detectors, including the photodetectors that are used in the
EPRB experiments, and he is the "father" of the quantum cascade
laser. He knew about EPRB experiments and the Bell theorem. He

Einstein Was Right!
Karl Hess
Copyright © 2015 Pan Stanford Publishing Pte. Ltd.
ISBN 978-981-4463-69-0 (Hardcover), 978-981-4463-70-6 (eBook)
www.panstanford.com

also knew that I would not be calling him if I was not convinced about having important results. After presenting my case, I asked Federico whether he would, as a member of the National Academy of Sciences, sponsor a paper for *PNAS*, the proceedings of this National Academy and a very prestigious journal. I was at that time not yet a member of the academy myself. Any member of the academy could communicate annually a small number of papers to *PNAS*, determine the referees, and be involved in the publication process in the role of a guest editor. If the referee reports were good, such a paper would be printed with very high probability. For academy members, this was a wonderful way to publish their own research and also to sponsor deserving research of young scientists as well as research on somewhat risky but highly interesting topics.

The tradition of such a special treatment of academy members came from Europe, and one of the best known cases of an academy member dealing with risky work is that of Bose. Bose discovered a new way of dealing with the quantum statistics of photons and certain atoms. His beautiful paper was rejected, and he sent it in his despair to Einstein and asked whether Einstein would sponsor its publication. Einstein did indeed wish to sponsor it and translated the paper personally. He reported on Bose's paper in sessions of the Prussian academy and got Bose's work published. Einstein subsequently extended Bose's work, which resulted in well-known papers that still form a basis for a major portion of quantum statistics.

As an aside, this shows that Einstein was not against probability in quantum physics. On the contrary, he was one of the founders of the new quantum statistics and a master in statistics and probability theory, as anyone who reads his papers can see. Of course, I did not think that our paper would come close to the importance that Bose's paper had, but I considered it important enough to ask Federico to help us get it published. Federico responded like a good friend and great scientist should. He said, yes, he would start the process and would send our paper to two referees. I know now that the referees were Roland Omnes from France and Marlan Scully from Texas. When Omnes and Scully told me later that they were the referees, I felt like the horseman who rode over a large frozen lake without knowing. When he found out, he dropped dead from the horse, so the story goes. Omnes is and was at the time one of the world's top experts in quantum probability (see the "consistent history approach") and Scully one of the world's top experts in quantum optics. This was our first work in the area of the foundations of quantum theory, and this was a very severe scrutiny. Federico's decision was, of course, the right one. If our work could not withstand that scrutiny, it was not worthwhile to be published by PNAS. Fortunately, the referees liked the paper and particularly our dealing with the role of time. There was a little hitch. Our paper was too long, but the referees agreed that it could be split into two papers, one that introduced our physical idea about the role of time (and the time- and setting-dependent equipment parameters) (Hess and Philipp, 2001) and a second paper that presented the mathematical details of Walter's model (Hess and Philipp, 2001). This latter paper was based on Kolmogorov's probability theory and is, I believe, a very serious counterargument to Bell. Its essence has already been discussed above in Section 4.2. Both papers appeared in the December 2001 issue of *PNAS*. The interested reader is referred to these publications for the more detailed description of our work. Walter and I were happy that our papers were printed, and we were thrilled when Philip Ball of *Nature* discussed our work in a very nice article in *Nature News* titled "Exorcising Einstein's Spooks."

5.2 Shots from the Pulpit

What we had tried to exorcise was, of course, the spooky influences at a distance that were now strongly supported by Bell's work. Philip Ball gave a short description of the discussions of Bohr and Einstein, the EPR paper, and Bohr's response. In a few paragraphs of masterful journalism, he carved out that Einstein had given Bohr the choice of spooky action at a distance or an incompleteness of quantum mechanics because it did not deal with "hidden parameters (or variables)" that determined the single outcomes of the experiments. He explained that as soon as it became possible to perform EPR experiments, the results showed that hidden variables did not exist provided John Bell's theory was right. Ball then introduced our work and explained our attempt to show that for variables related to space-time Bell's theory breaks down. "They find that if hidden variables have properties that change over time, yet are related to each other, the predictions change. For example, the hands of a clock in London and a clock in New York ... are correlated with one another ... in such a case, the results of EPR experiments can be explained without needing to invoke the spooky action at a distance that Einstein considered so unlikely." At the end of his article Ball also pointed to two preprints of experts who thought that Walter and I "have not unequivocally escaped the strictures of Bell's theory." These two reprints were not available to us at the time. We did have extensive information on and perhaps knew the content of the preprint N. David Mermin's, because we had e-mail correspondence with him. We did not have knowledge of the other preprint by R. D. Gill and colleagues, although we also had e-mail exchanges with Richard Gill. This latter preprint was a response to our paper in *PNAS*, but *PNAS* had not informed us about it, which we thought was quite unusual. There were also other papers published in opposition to our *PNAS* papers, and I am discussing these and Tony's response in some detail.

5.2.1 *Leggett*

We had shown Tony several drafts of manuscripts that explained our ideas about the role of time. His objections ranged from comments such as "You must have missed some subtlety" to little red wavy lines on our manuscript. Tony repeatedly told us that, while a measurement was performed with setting **a**, he could instead have turned the setting to **c**. Then one would have obtained the result for setting **c** at that same time. This was his main argument for using the same λ in every term of Eq. (3.1). "How does Tony know that no sequence of our time-dependent tables exists that works for all possible choices of settings?" was Walter's question. I thought I knew the answer to that. Tony was thinking of the limit of having at each point in time all the different settings, as illustrated in Fig. (4.3). We were back to the restaurant "possessing" all the food and customers eating only select choices.

Tony also explained to me again that the absolute randomness of the choices of settings was very important for Bell's argument.

He emphasized that the experimenter in each wing had free will to choose any setting at any time and therefore random setting pairs could be chosen at random times. His recourse to "free will" in the explanation of a physics problem was, as mentioned previously, suspect to me. I knew from my Monte Carlo modeling experience that a free will to turn knobs into arbitrary positions is an illusion; every setting and turning of knobs is subject to the laws of physics. As mentioned, the equipment settings are by their nature macroscopic. This means that changing them takes a finite amount of time, because massive sets of particles need to be shifted. This is true for the turning of a polarizer, even if it is done electrically. Electrical charges controlling polarizer settings require changes of macroscopic voltages and therefore involve the transport of many electrons. According to the theory of relativity, such changes take a certain time, because nothing moves faster than light in vacuum. The changes are also related to a variety of dynamical effects of the particles constituting the measurement equipment, and these dynamical effects are not all controlled by the experimenter. It is thus not possible for the experimenter to choose at any point of time any setting with given configuration of its constituent particles (electrons, atoms) and fields (e.g., electrical).

Another important fact is that once a setting (**a**, **b**, or **c**) is chosen, no other setting can be chosen at that time or space-time coordinate. It is, therefore, possible that a certain λ that is emitted from the source at one particular time can be linked only to one setting, the setting that was chosen just at that time. Therefore, Bell's choice of equal λs for three pairs of experiments is a major assumption that does not have to be true. There was, however, Tony's remark that he could have chosen different settings and then measured with the same λ and thus the particle "possesses" all these other outcomes. To satisfy these concerns, we constructed a more detailed restaurant counterexample, mostly to satisfy our own concerns.

Consider restaurants and guests on the two islands Tenerife and La Palma that can be described as follows. On Tenerife, guest G_1 dines in an Indian restaurant that has its menus labeled $A_{\mathbf{a}}^{st_i}(\lambda_i)$ as well as in a French restaurant with menus $A_{\mathbf{b}}^{st_i}(\lambda_i)$. On La Palma, we have guest G_2 dining in another French restaurant with menus

$A_{\mathbf{b}}^{st'_i}(\lambda_i)$ and an English restaurant with menus labeled $A_{\mathbf{c}}^{st'_i}(\lambda_i)$. Every day these restaurants pride themselves on using different spices, or at least different amounts of spices, which is reflected in the space-time superscripts of our functions (or just the index i that numbers the days). All restaurants use common meat and fish deliveries that are indicated by the variable λ_i and may change every day also. Each guest is permitted to eat in precisely one restaurant per evening on their respective island. There are clearly correlations between the menus of the two islands, because Indian and English dishes are often related to each other by the use of curry and because of the same meat and fish source. We can now (1) take an average of the stomach contents of guests G_1 and G_2 over long periods of time and (2) average all the menus of the restaurants. The question is, then, are the averages (1) and (2) the same? The answer is that in general they are different. The use of spices is neither independent of time nor independent of the settings (English, French, Indian). There is even a question whether the average spicing that is sampled by a given guest over time is equal to the average over all menus of the restaurants (all they "possessed") that the guest visits. Therefore, the mere fact of "possessing elements of reality" does not let one predict what can be said about actual measurements and what the long-term expectation value of these will be. Our conclusion was, therefore, that Tony's assumption that only one of the terms in Bell's inequality needs to be measured and the other terms can be inferred and added up is generally incorrect. Mathematical details, related to what is actually "sampled," will be discussed in Section 9.3. Trusting Walter with his restaurant example was in hindsight the right thing to do and was also important to deal with other scientists' objections to our work, particularly those of N. David Mermin.

I also made, however, an unfortunate misjudgment. Walter noticed immediately that a sequence of a pair of functions (or apps) such as $(A_{\mathbf{a}}^{st_i}(\lambda_i), A_{\mathbf{b}}^{st'_i}(\lambda_i))$ could under given circumstances be what mathematicians call a "vector stochastic process," and he said that he would formulate his model in the language of such stochastic processes. I said, "Walter, do not do this. Physicists know stochastic processes from Brownian motion, and Bell must have known stochastic processes also. I do not think that it's possible to refute Bell by just invoking Brownian motion." "Well," said Walter, "I could go to Joseph Doob and ask him whether he thinks that Bell's proof can be executed for stochastic processes." Doob was the world's top expert for the mathematics of stochastic processes, the "father" of the so-called Martingales, and he was Walter's friend. They were often hiking together, at least when Doob was still a lot younger. Both of them could eat copious amounts of hikers' delight, a mixture of bacon and onions that would have coated the arteries of normal persons instantaneously. "Asking does not hurt," I said, and Walter did ask. Doob told him, as far as I remember, that he did not think that the Bell proof could be completed for stochastic processes the way Bell did it. However, Doob did not want to strain himself by any of the detailed physics related to influences at a distance. He probably also did not wish to get involved and tangle with Tony. So, I finally asked Walter not to emphasize stochastic processes directly, because of my fear that Bell somehow must have known how to refute objections based on something like Brownian motion. I think now that this may have been a mistake. Stochastic processes can describe much more general physical processes than Brownian motion. My prejudice came from a lack of knowledge of stochastic processes, and we could probably have presented our case more crisply by referring our expert readers to vector stochastic processes of a very general form.

5.2.2 *Mermin*

David Mermin did have an initial positive reaction to our ideas about the involvement of space-time, as he later admitted in one of his e-mails. He said that he told his wife, "Something wonderful has happened," meaning that we had laid a big egg. However, very soon his e-mails became negative and he gave us the standard arguments on, as we knew from Tony, why we could not be right. After our *PNAS* paper appeared, he told me in an e-mail: "Karl, you have made a spectacle out of yourself."

Mermin (Mermin, 2002) did not consider any of our detailed mathematical treatments of time-related variables. As was typical for most reactions to our paper, he presented another "proof" of Bell's theorem that he had come up with previously. It was one of the very specialized examples based on a few "instruction" sets that the entangled pairs must "possess" if they cause the outcomes of the measurements. Mermin assumed possible settings **a**, **b**, and **c** that could be randomly chosen for both wings of the experiment (on both islands). After each choice of setting, a measurement was made resulting in green (for the result +1) or red (for the result −1) flashes of a hypothetical measurement machinery. From these facts, Mermin deduced eight instruction sets that each particle that is emitted from the source must have in order to "possess" the outcomes for all settings. These eight instruction sets are RRR, RRG, RGG, GGG, GGR, GRR, GRG, and RGR. The meaning of these instructions is that R stands for red and G for green flashes, respectively, and the first letter denotes the color for setting **a**, the second for setting **b**, and the third for setting **c**. Mermin (Mermin, 2002) outlined then that quantum theory demands the following results:

  (i) "In those runs in which the detectors happen to have been given the same settings, the light always flash the same color."

 (ii) "If all runs are examined without reference to the settings of the detectors, the pattern of flashes is completely random; in particular, the colors flashed are equally likely to be the same or different."

Subsequently, Mermin showed, by using methods of proof similar to those of Bell and Tony (but now simplified to eight instruction sets instead of a general λ), that statement (ii) cannot be achieved. He then stated, "Confusion buried deep in the formalism of very general critiques tends to rise to the surface and reveal itself when such critiques are reduced to the language of my very elementary example." Mermin's example does indeed highlight the crux of the problems when dealing with questions related to Bell's paper: Bell is always right when the elements of reality are like colored marbles (or like Mermin's instruction sets) and if the equipment settings have no dynamics and can be represented by the constants **a**, **b**, **c** .... Such elements of reality are, however, not general enough, and elements of reality related to space-time measurements need also be included as we have shown.

Walter looked up from Mermin's manuscript with a tortured face and said, "He wishes to restrict our functions of time to eight instruction sets! He claims that our model 'collapses' onto his; this is mathematical magic. He does not even realize that both our functions and λ can change depending on the time of measurement." However, it took Walter several weeks to come up with the detailed algebra that proved to ourselves that Mermin's criticism was unjustified. Here are our reasons why we thought Mermin's response did not hold water.

Mermin wrote: "It is also worth remarking that according to quantum mechanics the statistical character of the data in an EPR experiment is unaffected if the two detections are separated by arbitrary long-time intervals." This is true. However, the quantum result is conditional to the fact that the correlated pairs are actually measured. To achieve this, certain space-time correlations need to be fulfilled for all measurements, because the correlated pairs propagate with a given velocity; for photons this is the velocity of light. If one measures photons in one experimental wing with a time delay of 1 second, then that measurement needs to be performed at a spatial distance comparable to that to the moon, because light travels in about 1 second to the moon. Our functions of space-time (or the apps that we discussed) can indeed achieve equal "statistical character of the data" for any sequence of space-time coordinates. It makes no difference for our apps whether the measurements are performed this moment here on earth or a second later on the moon. The apps can depend suitably on the space-time coordinates.

Mermin further comments that to maintain this feature the Hess–Philipp micro-settings need to be the same for all time. This comment shows the attempt to eliminate space-time by making everything the same for all times. Mermin demotes our special introduction of space-time by throwing space-time together with the settings to form his "micro-settings." Thus the actual setting together with the time of measurement becomes a micro-setting. This micro-setting is then randomly picked. Time is thus randomized. In addition, Mermin's eight instruction sets that replace λ cannot exhibit any time dependence, because they occur randomly also, just as the random micro-settings do. Mermin then treats the micro-settings and his eight instruction sets, and this is again the crucial point, as independent mathematical variables. The λs are thus replaced by eight instruction sets representing one mathematical variable, while the micro-settings represent a second variable and both variables are treated as mathematically independent. The ordering or labeling role of space-time and any possible dynamics of the particle–instrument interactions are thus ignored.

Mermin's "crisp" example is, therefore, oversimplified and lacks the necessary generality to be applied to the EPRB experiments. It is also experimentally very difficult to realize. The reason is the relatively large experimental error that is typical for EPRB experiments. This large experimental error, unusual for other types of quantum experiments, will be discussed in Section 13.6. The error bars of the well-documented experiment of Weihs and coworkers (Zeilinger group) are close to 11% for measurements with equal settings. With such a large error, one can easily come close to satisfy condition (ii) and can definitely violate Bell's inequality.

5.2.3 *Gill, Weihs, Zeilinger, Zukowski*

I do not remember when I received the first e-mails from Richard Gill. He had another model for the functions A of Bell, very similar to that presented by Tony in his seminar. It was again based on the assumption that all data could be ordered into sets as shown in Eq. (2.2) with identical λs. Once such an assumption is made, the Bell inequality must be valid. Thus Gill assumed what he had to prove. There was a subtle difference to Tony's presentation. Gill, at first, did not state that only one of the terms of Eq. (2.2) needed to be measured and the others were included, because the quantum particles "possessed" such outcomes. He considered a large number of simultaneous two-wing EPRB experiments and let me know by e-mail that he could perform such experiments simultaneously at millions of places all over the Milky Way. Therefore, he argued, he did not need to consider different times. I did mention to Gill that such simultaneous measurements could not be performed in different inertial systems of the galaxy according to Einstein's theory of relativity, because all these experimental arrangements had different space-time coordinates and what appears as timelike in some solar systems of the Milky Way may have also spacelike properties in others. The outcomes of the experiments could now also depend on the space-time coordinates that were all different for the many different experiments. Walter also gave Gill his example with the restaurants and that one could not argue to have all equal times in Eq. (2.2). We did not get anywhere with Richard, and I got so annoyed that I instructed my PC to reject his e-mails. The *PNAS* publication of Gill, Zeilinger, and others (Gill et al., 2002) reiterated some of Gill's comments.

Gill's argument about measurements all over the galaxy was included in their publication, where they wrote: "Consider (as a thought experiment) repeating the measurement procedure just described, not as a sequence of successive repetitions at the same locations, but in a million laboratories all over the galaxy." I do not understand how Gill's coauthor, the great experimental physicist Anton Zeilinger, could suggest the possibility of such simultaneous measurements. Their paper also stated: "We did not mention time in our derivation because it was completely irrelevant. We did not compare actual outcomes under different settings at different times, but potential outcomes under different settings at the same time." This was the argument along Tony's lines of "possessing" properties, and we were by now convinced that we could refute this argument.

Thus Gill and coworkers declared time and space-time as "completely irrelevant." However, influences at a distance, that Anton Zeilinger seems to have approved, cannot be considered without using the space-time concept either. To define influences at a distance, one needs to assert that one measurement at a certain space-time coordinate st_i influences instantaneously another measurement at a second space-time coordinate st_i'. The expert of relativity theory notes that such a definition cannot be made consistently for objects and observers moving with different velocities. Thus space-time is not irrelevant for Gill and coworkers and certainly not for the Aspect–Zeilinger type of EPRB experiments. In fact all of these types of EPRB experiments use clocks to register the correlation within a certain time interval, the time window. If space-time were indeed irrelevant as claimed by (Gill et al., 2002), it also must be irrelevant to the experiments. What one needs to show, then, to exclude space-time is the existence of an EPRB experiment that violates a Bell inequality and does not use any clocks for the measurements but still excludes information exchange with the speed of light or lower speeds. The trap that Einstein has set becomes now very clear. He suggested an experiment with correlations at two different locations. How can one perform and describe such an experiment without the use of space-time? As soon as one removes the use of clocks and/or distance measurements in such experiments one operates in a logical vacuum. It is impossible to even imagine how a complete theory could describe this situation of correlating distant events without the use of any concepts related to space-time. It is equally impossible to imagine how to perform such an experiment without clocks.

Moreover, Larsson and Gill later showed explicitly that the Bell inequality is inadequate for experiments that use a time coincidence technique (time window) to identify pairs of photons (Larsson and Gill, 2004). They gave a specific example for time- and setting-dependent parameters, parameters that Walter and I had proposed in all generality in our 2001 *PNAS* paper (Hess and Philipp, 2001). Of course, they did not recognize the validity of our second *PNAS* paper (Hess and Philipp, 2001) and claimed that our model was "not local in the Bell sense." As "local in the Bell sense," they defined a model with a probability measure independent of the settings. This definition makes, in general, no sense, because the products $A_a A_b$, $A_a A_c$, and $A_b A_c$ and their long-term average must depend on both settings. Their definition also has nothing to do with Einstein's locality that follows from the limiting velocity of light. This is the only locality that matters. Furthermore, Larsson and Gill stated now "there is no problem in using the measurement time." Thus they did not realize that the inclusion of time makes it impossible to treat the settings as additional independent mathematical variable. I will return to the difficulty of including time in Bell-type proofs in the next section related to Myrvold's response. The results of Larsson and Gill were in their essence already found previously (Pascazio, 1986) and can be summarized by the following delay-time model that clearly demonstrates the Achilles' heel of Bell-type reasoning.

Assume that the measurements are performed with detection instruments that are mounted at different distances from the source. Let the detectors in Tenerife and La Palma be located in such a way that the measurement in Tenerife with setting **b** is performed

5 nanoseconds earlier than all the other measurements in both Tenerife and La Palma. This can be achieved by putting the **b** setting polarizer 5 foot closer to the photon source. According to Mermin, the quantum result for the statistical correlations is unaffected if the detections are separated by time intervals. Thus we expect the same results for all entangled pairs. Yet, this is not what one measures if one adds the just mentioned time delay of 5 nanoseconds and uses a fixed coincidence time-window of say 10 nanoseconds. Then the polarizers with settings **b** in Tenerife and **c** in La Palma give a coincidence only if the particle in Tenerife arrives in the first 5 nanoseconds of the coincidence window. For later arriving particles, the particle in La Palma arrives outside the coincidence window and is not registered. Because we are used by now to the intricacies of Bell's inequality, we note immediately the following. If we have $A_{\mathbf{a}} = A_{\mathbf{b}} = A_{\mathbf{c}} = +1$, then the first two terms of the Bell inequality result in $A_{\mathbf{a}}A_{\mathbf{b}} + A_{\mathbf{a}}A_{\mathbf{c}} = 2$ but the third term needs to be deducted, which reduces the result by 1 and makes the Bell inequality valid. If, however, the third term with the setting pair **b**, **c** occurs only half the time in the coincidence count as compared to the other setting pairs because of the delay, then we deduct less than 1 (e.g., 0.5) and the Bell inequality is violated. Such examples can sometimes be refuted, for example by special experimental provisions that make such delay impossible. However, if one just reduces as a Gedanken experiment the time window and delay time by factors of 10, 100, 1000, and so on, one can see that refutations become more and more difficult. Walter and I did appreciate that it was not wise to present such specific examples in our *PNAS* papers, because specific examples can be circumvented by specific measurement arrangements. However, the general setting- and time-dependent equipment parameters that we introduced open the floodgates.

I will show later, when discussing the work of Boole and my collaboration with Hans De Raedt and Kristel Michielsen in Chapter 13.2, that the argument involving Einstein local space-time related effects can be made in such a general way that there is always room to refute Bell's proof for any EPRB experiment.

5.2.4 *Myrvold*

Myrvold (Myrvold, 2002) tried to refute our paper in a different way. He maintained, on the basis of previous publications, that Bell implied two forms of "independence" in his work. One was setting independence: any given outcome in one city, say Tenerife, cannot be influenced by the setting chosen for the given entangled pair in La Palma. This fact expresses Einstein's views of locality based on the limitation of all velocities to that of light in vacuum or lower. Without any instantaneous influences at a distance (nonlocality), no communication is possible between the measurements of the given pair in the two distant wings of the experiment. Walter and I agreed, of course, that this is a true fact.

The second form of independence implied by Bell, according to Myrvold and others, is the so-called outcome independence conditional to λ. Outcome independence conditional to λ represents a concept introduced by Bell and means that for any given λ the outcome in one wing is independent of that in the other. Naturally, if λ represents just marblelike entities (as opposed to space-time), this needs to be true, because Bell's functions A depend only on λ and on the settings. However, it is also true that Bell's λ cannot include a space-time coordinate, else we cannot regard the A's as functions of two independent mathematical variables as Bell does. One therefore needs to introduce, as we did, a separate space-time index. Then, the outcomes conditional to λ may still depend on the space-time coordinates, and outcome independence conditional to λ is not guaranteed. For example, clocks in the two stations can show correlations no matter what was sent out from a source.

The problem gets worse if one claims, as Bell, Tony, Myrvold, and others, did that λ can be "anything." Here one can see the pitfalls of the use of new concepts together with generalizations. Outcome independence means, then, independence conditional to "anything" and therefore means complete independence. Naturally, one cannot describe correlated events on two islands by any theory that demands complete statistical independence. The new concept outcome independence conditional to λ and the generalization of λ to "anything" have formed a vicious logical circle. We will see later, in connection with Boole's work, that this is not the only logical circle in Bell-type theories. I believe the discussions of Bell's theory have been obfuscated by the introduction of mathematically ill defined terms such as "the free-will choice of settings," "particles possess," or "outcome independence conditional to elements of reality," which are then applied in connection with generalizations, such as a λ, that can be "anything" or a setting that is independent of "anything." Kolmogorov's probability theory, as discussed in Section 7.3.2 defines conditional probabilities within an "algebra of events," which is the only way of rigorous definition. Obviously, one cannot take "anything" and hope that it follows the laws of algebra. We will return to this point in more detail in Section 9.1.

5.3 Strategy for Further Investigations

After a careful reading of the papers by Gill et al., Mermin, and Myrvold, I suggested to Walter that we work on a refutation for all known special Bell-type proofs. Fortunately, most of these proof types could be found in Bell's collected works (Bell, 2001). Walter did not like this idea at first and it took a lot of coaxing to get him going, but this is what we did during the following two years. I also talked to Tony about the papers of Mermin and Gill et al. Tony was very informed about these papers and could not be persuaded that we had a point against them.

The back and forward of our discussions ended in a challenge that Tony put like this: "Show me that you can play the Bell game!" Have two players in two separate measurement stations choose outcomes for the functions A without knowing anything about each other. The settings in each city should be chosen randomly by someone and each of the players knows their precise setting at the time of measurement. For each given pair of settings, a decision of each of the players is required to provide a value of $+1$ or -1. The players have no clue of each others choices. They are, however, permitted to know all results from past experiments. A further idealization of Tony's game was that the perfect detection of all entangled pairs was assumed. It had to be played as if each player made a measurement with a guaranteed random setting and the additional guarantee that one part of an entangled pair was measured; no influences of the equipment such as absorption of the photon or time delay of detection permitted. As far as I understand it, it was also at least implicitly assumed that for a given clock time there was exactly one detection in each station and it was guaranteed that an entangled pair was detected.

"If you can come up with choices that lead to a violation of a Bell inequality, then you win," said Tony. This is what, according to Tony, nature accomplishes in the EPR experiments—it provides us with two outcomes in measurement stations that, if Einstein is right, do not "know" about each other at that moment. What Tony did not mention, of course, was that nature does not "know" whether or not an entangled pair is indeed measured and nature does not "know" any coincidence of distant measurements. Observers moving relative to the EPRB experimenters would observe different coincidences. The concept of coincidence is entirely manmade and depends on a time window monitored by the experimenters clocks. If this fact is disregarded, as postulated by Tony, then winning the Bell game becomes very difficult.

Detailed considerations, particularly those discussed in connection with Boole's work in later sections, show that one can only hope to win the game in three ways. The first is covered in a later section when discussing the work of Pearle as well as De Raedt and Michielsen and relates to so called filters. The second is to find connections between the choices of the setting pairs and the measurement times and relates to the fact that the measurement may contain complicated time and setting dependent "machinery" based on many body effects in the polarizers and also the time window measurements. An example would be that measurements with settings **a**, **b**, and **c** are necessarily performed all during different time periods after a particle hits the measurement instruments, because different settings involve different many-body configurations and therefore different many-body interactions of the incoming quantum particles with those that constitute the measurement equipment. This approach requires the detailed knowledge of the machinery of measurements, including all possible many-particle interactions. The third possibility is to find an infinite set of functions A^{st_i} and $A^{st'_i}$ that return a violation in the long-time average for all possible setting choices. Walter and I saw immediately that this latter way was a tall order that required knowledge of all possible function outcomes for all possible settings at all possible times. The randomness of the polarizer settings on each side and equal outcomes for equal polarizer settings present, then, a big problem.

Our own model could not play the Bell game to start with, because our model required the knowledge of a probability for the measurement outcome, a so-called joint probability, which is a standard tool that makes probability theory tick. Any joint probability requires the knowledge of the settings in both wings, except for stochastically independent experiments. Thus Tony's rules for the Bell game excluded any standard probabilistic approach. Also, quantum theory requires the knowledge of both settings to calculate the average over many EPRB measurements. Therefore, the Bell game as formulated by Tony cannot be played with quantum theory either. Tony did, of course, put his finger on a serious question. How could we be sure that our time-dependent probabilities could indeed be generated for any random setting sequence by time-dependent tables such as those given in Fig. (4.1)? All we had shown was that we could find tables that did the trick for any given and known sequence of setting pairs. That knowledge was seen by Tony as an illegitimate factor that may contain effects that are not strictly local. We will return to this important point in Section 13.1.

The strategy that I proposed to Walter was, therefore, to show in detail the deficiencies of every single one of the existing Bell-type proofs. To say it upfront, the basis for these refutations was always the hypothesis of the involvement of space-time. This involvement of space-time in the functions A makes it impossible to claim independence conditional to λ, because of the possibility of correlations due to the additional space-time-dependent effects. One can also not claim independence conditional to any space-time coordinate, because that would mean that the experiments in the two wings are completely independent and uncorrelated, which they are not. Walter asked if I thought that this effort was really worthwhile. He also was concerned about comments of some of our colleagues that quantum teleportation and quantum computing were sure to become very important. There was a paper from 1998 that seemed to verify the possibility of quantum teleportation that was proposed originally by Bennet and coworkers (Bennett et al., 1993). Walter was pointed to the news about teleportation by several of his colleagues and wanted to know whether these news proved instantaneous actions at a distance. I, therefore, did a very careful review of the teleportation literature and describe my findings in the next chapter. My conclusion was that quantum teleportation had only then substance if violations of Bell's inequality indeed indicated in the first place some instantaneous influences at a distance.

Chapter 6

Teleportation and Quantum Computing

Magnas it fama per urbes . . . monstrum horrendum ingens.

Virgil, *Aeneid*

Fame walks through big cities . . . a horrible monster, still growing.

The stories about teleportation did sound to me like a horrible monster that was growing while walking through the cities. Everyone who knew that Walter and I were opposed to the idea of instantaneous influences at a distance told us that quantum teleportation was now proven and that it demonstrated clearly the existence of influences at a distance. My usual response was that quantum theory had some very great successes with predicting the precise energies of electrons in atoms. The experiments relating to quantum information such as the EPRB-type experiments and the so-called teleportation did not agree with quantum theory with such precision but only approximately. Furthermore, as mentioned, I soon convinced myself that the idea of quantum teleportation assumes that Bell is correct and influences at a distance do exist. We thus have another logical circle. Quantum teleportation is only then something remarkable if influences at a distance are assumed to exist on the basis of Bell's work to start with. Experiments are explained using

Einstein Was Right!
Karl Hess
Copyright © 2015 Pan Stanford Publishing Pte. Ltd.
ISBN 978-981-4463-69-0 (Hardcover), 978-981-4463-70-6 (eBook)
www.panstanford.com

this assumption and then, because of a very approximate agreement of the experiments with the quantum predictions, it is claimed that teleportation has been proven. Here are the actual facts starting with the concept of a quantum computer.

6.1 The Quantum Computer

Richard Feynman proposed that quantum computers, computers that use quantum particles and measurements of them, might be very powerful machines that can solve complicated quantum mechanical problems in a natural way, because they use the quantum particles themselves as part of a big computational machinery, the "quantum computer." One example would be the experimental determination of the atomic spectra of electrons and molecules that can be measured to many digits while the corresponding computation takes the largest supercomputers, at least when complex molecules are involved. The question is, of course, whether one can use the measurements on quantum particles for more general purposes. Take, as another example, a strand of DNA and let it bounce on a diamond surface. Then measure its geometrical arrangement. It is easy to measure this geometry with an atomic force microscope and almost impossible to calculate all possible results. But even if one can calculate the possible results, what does one do with them?

Elaborate theories have been developed subsequent to Feynman's suggestion. It was shown that quantum computers do offer great opportunities, particularly if Einstein was wrong and influences at a distance exist. Then, and only then, quantum computing adds another "dimension" of advantage to conventional computing. As it turns out, it is very difficult to build such a machine for a variety of reasons and no quantum computer of any significant capability has been built yet. Many engineers and scientists believe that such a machine cannot be built at all, while others are more optimistic and believe that the difficulties can be overcome. The author knows that science and engineering have provided us with digital computers that far exceed the power that had been thought impossible to achieve just 50 years ago. The humble computational engines that started the computer age may have been as primitive as current quantum computer prototypes are. There also exist quantum devices with astounding new properties (such as superconducting quantum interference devices). I am, therefore, not attempting to predict whether or not it is possible to build such a machine and concentrate only on the second requirement for quantum computation, that instantaneous influences do indeed provide another "dimension" of advantages.

It is not generally appreciated that much of the predicted superiority of quantum computers, over what is known to be possible from the view of Claude Shannon's classical information theory, stems from the instantaneous influences at a distance. It is always emphasized in the quantum computer literature that information propagation faster than the speed of light is not possible. No signal that presents us with a 0 or a 1 can be transmitted faster than the speed of light. However, it is also always stated that entanglement is the main trademark of qubits, the bits of a quantum computer. The qubits are, in principle, nothing but entangled particles such as the photons of the EPRB experiment, and all the proponents of quantum computing insist that the qubits are influenced instantaneously by measurements or other operations on the entangled partners. This entanglement of many qubits leads to the possibility of a massive parallel interaction that does provide significant advantages, even though the speed limitations for the actual information are not broken. Only in this way does the quantum computer become very special. For, if Einstein is right and the entangled particles are just correlated by elements of physical reality, then the qubit operations are nothing else but the combination of so-called digital and analog computing.

The ordinary digital element of a conventional computer operation, the bit, is just a 0 or a 1, a current or no current, a voltage or no voltage, etc. In addition, it has been known already before the digital age that one also can use values of currents or voltages other than 0 or 1, continuously changeable aspects of physical phenomena, in order to perform computations. Computational engines that do this are called analog computers. Naturally, combinations of analog and digital computing are also possible. Naval gunfire computers of the past were of this kind, and some may still be in use.

The qubit of a quantum computer performs a combination of analog and digital computing in the following sense. The quantum outcome of a measurement is always digital and can be denoted by 0 and 1, or by $+1$ and -1, as we did for the EPRB experiments. This outcome occurs with a certain quantum probability that can assume all values of real numbers between 0 and 1 and thus represents a continuously changeable aspect, the analog computing component. If this analog component is based on an element of physical reality, as Einstein claims, then the quantum computer is nothing but a combination of digital and analog computers. If, however, Bell is correct, and there are no elements of reality that can describe the behavior of the entangled pair, if all these entangled pairs are influenced instantaneously wherever they are, then the quantum computer is a special machine that performs an instantaneous parallel operation of all the entangled pairs that are at work. It was this fact that attracted my interest and attention to Bell's work, because of my engineering background, and this is why I listened very carefully to Tony's lecture and worked with Walter and still work on and think about this topic.

Naturally, it is possible that certain quantum experiments may still provide an advantage, even if both the digital and analog computation can be described by Einstein's elements of reality and without instantaneous influences. Superconducting quantum interference devices, as mentioned, have very interesting and special properties that may lend themselves to innovative applications. However, the magic of massive parallelism that makes the idea of a quantum computer so attractive falls and stands with the existence of instantaneous influences at a distance.

These thoughts are best explained by using the example of quantum teleportation of a qubit. The teleportation that has been proposed by quantum theorists and later experimentally implemented (search the Internet for further information), shows clearly why many people think that the qubit is special and why it may be very useful. The teleportation process also explains the difference between instantaneous influences and direct information transfer that is never instantaneous. The use of a qubit in a "teleportation machine," a "teleporter," is therefore described next.

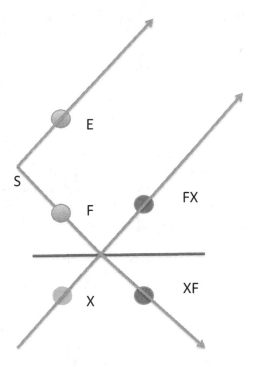

Figure 6.1   The teleporter that teleports the state X by use of the entangled photon pair (E, F) and a semipermeable mirror (horizontal line).

6.2 Teleportation

Figure (6.1) shows the basic principle of quantum teleportation as it was first introduced by Charles H. Bennett and his coworkers (Bennett et al., 1993). They use an entangled pair (E, F) as basis for their teleporter. The entangled pair of the later realized teleportation experiments is a pair of photons as it was and is used by the Zeilinger and Aspect groups.

In addition to the EPRB source that emits the two entangled photons, the teleporter contains only one additional, very simple device: a metal-coated glass plate. The metal layer is so thin that it transmits photons that are incident either from above or from below under a 45 degree angle, with a 50% likelihood. One of the entangled photons, F, is incident on the thin metal layer from above. Another independent photon, X, is separately prepared to be $+1$ polarized and hits the metal layer from below. If, as a result, two photons propagate toward the right as indicated in Fig. (6.1), then quantum theory teaches us that we cannot say whether F and X were both transmitted or both reflected and we cannot say which is which after they have hit the metal and propagated to the right. Therefore, in the language of quantum theory, the two photons propagating toward the right, one above (named FX) and one below (named (XF) the plate, are now entangled. This means, again according to quantum theory, that all of the three photons E, FX, and XF are entangled. Furthermore, if someone performs a measurement on the photon XF and finds that it is polarized $+1$ for a given polarizer setting, then quantum theory predicts that a measurement of E will give the same result.

This experiment has indeed been performed and the outcome has been confirmed with good accuracy, by now over considerable distances (e.g., the distance between Tenerife and La Palma). What has this experiment to do with teleportation? Bennett and coworkers explain that the entangled particles (E, F) carry special quantum information, *because Bell's inequality is violated in EPRB experiments and Einstein's elements of reality can, according to Bell, not exist.* Therefore, according to Bennett and coworkers, the quantum correlation is instantaneously established as soon as the polarization of XF is measured and does represent an instantaneous influence at a distance. Physicists perform such experiments and measurements, very expensive experiments, over longer and longer distances. Why does this not violate the generally acknowledged fact that information cannot be transferred faster than the speed of light in vacuum? Bennet and coworkers have the following explanation.

There are two experimenters named Alice and Bob. Notice that whenever these names are used for quantum information examples, we are talking about quantum theory experts that know everything about quantum mechanics. Alice attends the station with the very thin metal plate and makes sure that the X photon (that is prepared to be $+1$ polarized) and the F photon hit the plate simultaneously and propagate to the right, now as photons XF and FX. Alice also performs a measurement of the polarization on the XF photon. If the result is $+1$ (-1), then she knows that the corresponding E photon is also polarized in the +1 (−1) direction. Thus she knows that according to Bell she has "teleported" the +1 state of the X photon to Bob, the recipient of E. Had Alice prepared the X photon in the −1 polarization, Bob would now have (instantaneously) a −1 polarized photon. Why did we not transfer true information instantaneously? The reason is that Bob has no idea which photon Alice has entangled until she tells him, for example by cell phone or radio communication. But then, as soon as Alice has given Bob that information, Bob becomes the "master" of the additional quantum-type information and can use his qubit E accordingly. We thus see that we have not violated Einstein's ground rule that no information can be transferred faster than the speed of light, because Alice had to use a cell phone or something similar. But according to Bell, Alice has transferred, instantaneously, the +1 (−1) quantum polarization state of the X photon as soon as she performed her measurement.

Does that help us to do something useful? Not in the case just discussed. Alice could have told Bob on the cell phone also what type of photon he was receiving. However, if we have many qubits and they all communicate instantaneously their quantum polarization state, and thus act together, then a lot of smart scientists have shown that you can indeed perform enormous computer tasks based on these instantaneous interactions. That would be really great if it can be done. Now, however, consider the above experiment and assume that Bell is wrong and Einstein is right and we do have elements of reality that determine the outcome of all the polarization measurements.

From Einstein's point of view, there are no instantaneous influences at a distance. Elements of reality that can be carried by the particles and other elements of reality that describe the measurement equipment-dynamics, or even the environment, are responsible for all the information that is involved in the experiment. Anything that is done to the particles X, F, FX, and XF has no instantaneous influence on the distant E, which, therefore, does not change by Alice's machinations. If a +1 polarization is measured for XF, then Bob's E would have had the same polarization anyway. If Bob needed a particle polarized like X, why did Alice not send him X in the first place? The whole procedure with the thin metal layer acting like a semipermeable mirror is, according to Einstein, nothing but hocus-pocus, smoke and mirrors, as far as E is concerned. If Einstein is right, and this author has no doubt that he is, then a quantum computer is nothing but a combination of analog and digital computational elements, with the speed of light being the limit for all influences. That does not mean a quantum computer has nothing special to offer. As mentioned, superconducting quantum interference devices have very special analog properties even if one does not invoke influences at a distance. This author does not wish to say that quantum experiments do not offer advantages for computing. It is only the "magic" advantages of instantaneous interactions that I am opposed to.

The theoreticians working on quantum information-transfer add usually another assumption that would open new horizons, particularly for cryptography and the desire to transmit information by unbreakable codes. This is the assumption that you can carry the single particles (portions) of entangled qubits with you, so to speak in your pocket. Both Alice and Bob can then carry a large number of entangled qubits with them and, therefore, can release a lot of quantum information instantaneously to each other, wherever they are located. I believe that this possibility will remain forever a dream, because qubits interact with the environment and thus lose or change their quantum information. It is even difficult to see how the macroscopic polarizer settings can be assessed to be controllably the same on a rotating earth or in a satellite. I believe that it would be very important to first find out, with certainty, who was right, Einstein or Bell, and only if Bell comes out on top, highly unlikely in the author's opinion, one should spend large sums for quantum information and quantum computing based on instantaneous influences. Otherwise one needs to settle for more mundane advantages that may not be sensational enough to attract big funding.

Be all of this as it may, Walter and I were satisfied that we should continue our work that we hoped would bring about the decision that Einstein was right. We were also very encouraged by kind invitation letters that we received from two well-known scientists.

Chapter 7

Space-Time, Elements of Reality, and Probability Revisited

Denn eben dort wo die Begriffe fehlen,
da stellt ein Wort zur rechten Zeit sich ein.

—Goethe, *Faust* I

Free translation:

When concepts lack or spread confusion,
in time a word is offered as solution.

Soon after my investigations of teleportation, we received a very encouraging e-mail. Andrei Khrennikov invited us to attend a conference that he was organizing on the topic of foundations of probability and quantum mechanics and to present our work there. The conference was in Växjö, Sweden, and Walter and I started working on a major presentation.

I also realized that I was not really ready for discussions with experts at a conference, because my knowledge of both probability and quantum theory was too specific and exhibited black holes. I had read only a few introductions to the probability theory of Kolmogorov, which is the conventional probability theory most

Einstein Was Right!
Karl Hess
Copyright © 2015 Pan Stanford Publishing Pte. Ltd.
ISBN 978-981-4463-69-0 (Hardcover), 978-981-4463-70-6 (eBook)
www.panstanford.com

accepted by mathematicians. I did not really understand how it related to Einstein's elements of reality and Bell's λ, and I had only a very foggy idea about what the differences were between Kolmogorov's concept of probability and that of the fathers of quantum theory. I also had to do some learning to understand more modern views of quantum probability, particularly the view of Roland Omnes, who was the referee of our *PNAS* papers and was scheduled to attend the conference.

While Walter was busy refuting Mermin's objections, I asked him to discuss the foundations of probability theory with me. From these discussions I extracted the basic mathematical assumptions and their connection to the elements of reality of a physical theory based on a space-time approach. This was an important exercise for myself to gain reassurance and get prepared for discussions with other scientists.

There are many explanations for probability given in the literature. Some of them gave me reason to despair. One famous book starts by stating that one can talk about probability and work with it as long as one does not try to define the word. There are also discouraging words about space-time, such as the opinion of quantum theorists that space and time become just "foam" on the quantum level due to the Uncertainty Principle. Elements of reality were always thought of as colored and flavored marbles that do not change. Thus, the standard descriptions of these important concepts form a trinity giving rise to confusion, and reminded me of the three witches discussing when they should meet again. My attempts to sharpen and clarify the definitions of these concepts in generally understandable terms, yet with mathematical and physical precision, are described in what follows.

7.1 Space-Time

Locations and objects in space are determined by using a measure, such as a meter measure, and by recording the measurements within some system of coordinates, for example Cartesian coordinates (search the Internet for details). Coordinates are necessary for our interactions and discussions with other humans. In our daily life, we do not really need Cartesian coordinates that define the location of objects with precision. We may just say that our friend visited his aunt in New York for Christmas. This statement gives us the approximate location of the person on earth and also the time when the person actually was there. In general, one needs three spacelike coordinates and one timelike coordinate in order to interact with our fellow human beings in a logical way. As stated, the three spacelike Cartesian coordinates of objects can be measured by using a meter measure or the more complicated equipment of the global positioning system (GPS). The results of such measurements give us information about the spatial relation of objects to each other at a certain time that can also be measured, for example, by using a very precise atomic clock. The space-time data relate to the world and to possible changes, the dynamics, of this world at different clock readings, for instance the speed of a car as determined by GPS.

How important space and time coordinates are for statements about nature may be seen from the following example. Everyone knows that we talk about the "fixed" stars on the night sky as opposed to the moving planets and the moon. Yet, if we look at the night sky, all stars rise in the east and set in the west and thus move in front of our eyes, except for the North Star, because of the rotation of the earth. The stars are "fixed" only with respect to a coordinate system that compensates for the rotation of the earth. This example also tells us about the connections of space and time coordinates that, as Einstein showed, are so closely intertwined that one must talk about a space-time system, about a four-dimensional coordinate system.

The coordinates and clock times relating objects to each other, thus, form the basis for much of our logical reasoning. Events of nature are and need to be characterized by coordinates if we wish to discuss their dynamic existence and use value judgments such as true and false. For example, if we say the moon shines, that could be true or false depending on whether or not we give the location and time when the shining moon was observed. A cloud could have just moved by and covered the moon during some period of time. Some quantum theorists did seriously pose the question "Is the moon shining when I am not looking?" Einstein reportedly laughed at this question and found it silly. It is, of course, more precise to ask "Does the moon shine as seen from a certain frame of reference at a certain space-time coordinate?" This latter version opens the possibility that another person has looked, or an instrument in a certain reference frame (a laboratory) has recorded data on, whether or not the moon was shining. Such precision is necessary in scientific discussions. It is also the author's conviction that one needs to move away from ideas that the "conscious looking" of a person is necessary to assess whether or not the moon is shining. Precise reasons for this belief that is related to the concept of "macro realism" are given in Section 13.4.

Thus, the case can be and has been made for the absolute necessity of the use of space and time in science. The actual understanding of the concepts of space and time, however, have differed greatly over the centuries. Newton's work is, for example, totally based on considerations of the dynamics of massive objects in space, such as moon and planets and has regarded space as absolute and time as something that "flows" without any relation to any of the masses or their motion. No limiting speed existed for Newton, and he actually thought that the sun attracts the planets by instantaneous actions at the distance. Einstein's framework improved on this view of space and time. His relativity theory postulates a highest limiting velocity for all motion, which is the speed of light in vacuum and is usually denoted by the letter c.

This latter fact is used for our purposes in two ways. First, it is used to affirm the separation principle: whatever happens in Tenerife can only be influenced by any occurrences in La Palma with velocities smaller or equal to c. If we change the polarizer setting in La Palma just before one part of an entangled pair is measured, that measurement and polarizer setting in La Palma cannot have any influence on changing the measurement of the other portion of the entangled pair in Tenerife, and vice versa. Second, we can deduce from relativity theory that the polarizer settings can only be changed with a speed smaller than or equal to c, no matter how the polarizers are switched. It is, therefore, impossible to achieve different settings during the same small time interval, and one can only have one setting pair to measure one pair of particles. Furthermore, it is not possible to achieve random settings at random times continuously. If setting **a** is chosen, it takes some time to put **b** in place. Third, and this is very important, Einstein has shown that the concept of simultaneity of two events needs special consideration and that two events that are measured to be simultaneous by one experimenter may not be simultaneous for a second experimenter who moves with high velocity compared to the first. This fact is a problem for those talking about instantaneous influences at a distance, because the words instantaneous and simultaneous have similar connotations. If some events cannot be regarded as simultaneous at different locations, influences between these locations cannot be regarded as instantaneous without logical contradictions.

Instantaneous influences at a distance were anathema to Einstein. At the time of the EPR work, only *action* at a distance was discussed. Instantaneous actions at a distance, a direct instantaneous (faster than light in vacuum) transfer of information, were generally not thought to be possible and are still not considered possible by any serious physicist. Bohr and every member of the Copenhagen school did believe in this assertion of Einstein's relativity theory. Later, however, the word "actions" was replaced by "influences" with the following meaning. Because the single outcomes of the EPRB experiments are entirely random (the spin of a measurement is equally likely up (+1) or down (−1)), there occurs no instantaneous transfer of information. Therefore many quantum information theorists believe they can claim that the core of Einstein's postulates is not touched by this type of influences at a distance that they think happen for entangled particles. This twist of explanation from action to influence still contains fuzzy logic in the opinion of the author. The word "influence" implies a causal connection: if the spin is measured on Tenerife first and on La Palma just a tad later, as deduced from clocks fixed on the two islands, then that measurement in Tenerife is considered to "cause" the instantaneous influence on La Palma that results there in the measurement of the same spin. However, a fast-moving observer (relative to the islands and the measurement instruments) may observe the measurement in La Palma first and in Tenerife later, as we know from the special theory of relativity, and thus may conclude that the measurement in La Palma was actually the cause for the influence at a distance. If a large number of moving observers can all claim different measurements to be responsible for the outcome on the other side, these influences at a distance can certainly not be an element of a consistent logical theory. We are thus losing the logical basis for the statement that something happens instantaneously.

Einstein was aware of the deep significance of the thoughts just presented and also aimed to support his way of seeing things from a philosophical point of view. He realized that his views did not agree with the extreme proposition that all science should be exclusively based on direct sense impressions or data obtained from some machinery facilitating the sense impressions (Einstein, 1954). He gave the example of the system of natural numbers, the numbers we use when counting. The mathematical system of these numbers goes way beyond what one can derive from sense impressions, like the counting of oranges. For example, we use the fact that there exists no highest natural number and there exists an infinity of numbers. The mathematician Peano based the natural number system on axioms and derived mathematical truths, so-called theorems, that certainly cannot be obtained from raw sensory material. Thus, Einstein concluded that in order to do science we need to use elements other than the raw sensory material. He gave the example of two identical tall buildings, one in New York and one in Paris, and said that one would be, on the basis of raw sense impressions, forced to regard them as the same object in the sense of physics. He then stated that there was no danger (of logical wrongdoing) in combining the object as an independent concept with the proper spatio-temporal structure, that is with his space-time. He, therefore, combined the existing thing (the tall building) with all its qualities and takes the geometrical relations to other objects of the world as an additional quality. These geometrical relations involve space-time as a product and tool of our thinking that goes beyond the mere data of clocks and meter measures.

Einstein thus justified his use of space-time as an additional element of reality also philosophically, by saying in essence that we need to go beyond raw sense impressions if we wish to scientifically understand the world and the universe. He dealt with this subject with extreme care, because he was aware that we have here a very slippery slope. If we use such elements of human thought in addition to the elements of raw sense impressions, what are the limits for our theories? The quantum concept of superposition of states and Einstein's opposition to it that is described below are but one example of how carefully every new idea must be checked and discussed as soon as it transcends sense impressions. Modern theories of space and time are even more revolutionary and the reader is referred to Stephen Hawking's explanations of imaginary time (Hawking, 1988). It is worthwhile to note, however, that the still unsuccessful attempts of unification of quantum theory and Einstein's general relativity, in particular the work of Hawking related to black holes, demonstrate mathematically how quantum superposition (see Eq. (7.3) below) "clashes" with Einstein's relativity.

Fortunately for us, the time concept of probability theory needed to understand the following is simple and virtually identical to the concept of the natural numbers. The quality of time that is needed in probability theories is mostly its ordering character, its "ability" to order a sequence of events in some logical fashion and to assign a one-to-one correspondence between experiments and the mathematical abstractions that are used to describe them. This one-to-one correspondence and order is very important, as we will see later, particularly in connection with Boole's work. The interested reader may wish to search books of probability theory and will find that in all of them time is just represented by natural numbers as long as the events are countable. This is also the case for time-dependent stochastic processes and other frameworks of probability theory that are used to describe nature when some kind of randomness is involved in the outcomes of measurements.

7.2 Space-Time and Elements of Reality

7.2.1 *Einstein's Elements of Reality*

To understand Einstein's ideas of what elements of physical reality are, we need to keep the above facts in mind and learn about a few facts related to his space-time picture. Einstein was considering investigations of the four dimensions of space-time by involving measurement rods (such as a meter measure) and clocks that he imagined located everywhere in space while thinking of moving objects. For Einstein, the results of such measurements were as real as the existence of the moon was, although space-time data are man-made, while the moon just exists independent of us and was there before man populated the planet. Thus we have up to now two types of elements of reality that Einstein involved in his scientific thinking: (i) objects of nature detectable by us in some way such as the moon and elementary particles and (ii) human concepts, concepts of our thinking, as well as space-time measurement data that relate these objects to each other and address their dynamics. All these elements of reality can be represented by mathematical abstractions such as the functions A.

Einstein involved also a third type of elements of reality, a type that may not have yet been detected or recorded and may thus still be hidden to us and our knowledge. He deduced this third type of elements from correlations of measurements. For the case of the EPRB experiments, Einstein maintained that if one measures a correlated pair and if one can predict the outcome in Tenerife with certainty just by knowing the outcome in La Palma, then one can be sure that some element of physical reality is responsible for that correlation, even though that element is still not accessible to us in other ways.

Einstein, Podolsky, and Rosen stated: "If, without in any way disturbing a system, we can predict with certainty . . . the value of a physical quantity, then there exists an element of physical reality corresponding to this physical quantity." Consider, for example, two sticks that both rotate clockwise and are emitted from a common source, one toward Tenerife and the other toward La Palma. If the observers in Tenerife detect a clockwise rotation, they are certain that the observers in La Palma will also detect a clockwise rotation. Thus Einstein maintained that if some fact can be predicted from the measurement on one side with such certainty (mathematically called with probability one), then an element of physical reality (in the above example the rotation of the sticks) must be related to this fact. One can derive the existence of this element of reality even if it is "hidden" and unaccessible to our current measurement equipment and even if we deal with atomic spins and not just with the visible rotation of sticks. Naturally, the hidden elements could also be related to the measurement equipment. If we, on the other hand, accept the Copenhagen interpretation of quantum mechanics for the above example, then the spinning of the sticks is only "precipitated" in the moment of measurement and does not exist as a physical reality before that measurement is done. Then we have only one alternative to explain the correlated rotation of the sticks in the measurement outcomes and that is instantaneous influences at a distance.

7.2.2 *Bell's Elements of Reality*

Bell uses for his elements of reality the picture of things that can be drawn out of an urn. Of course, Bell knew that photons have both particle-like and wavelike properties, as do all other elementary particles or "entities." His assumptions show, however, that both his λ and the measurement equipment do not have general dynamic properties. The measurement equipment is explicitly described by settings $a, b, c \ldots$ that are randomly chosen but otherwise static, and general time dependencies contained in λ are excluded, because of the assumption that λ and the settings are independent mathematical variables. Space-time data, which are also elements of reality, go beyond the mere existence of particles (or wavicles, the entities of quantum mechanics that also exhibit wavelike properties) and provide us with the relations of these particles to each other and to an objective world. It is very important to realize that the description of the *dynamics* of the particles requires these additional man-made elements of reality.

Other descriptions of the objects of this world not using a space-time coordinate system are possible. As we have heard, we may describe a magnet just by the direction of its magnetic field, or a polarizer just by a direction that it points to. Such other descriptions, however, cannot be seen as independent from given Cartesian coordinates and clock times that describe the relation of all the elements of our surroundings to each other. There exists, therefore, a significant difference between elements of physical reality such as marbles in an urn, on one hand, and measurement times or space coordinates and the concept of space-time, on the other. Space-time and equipment settings cannot be regarded as independent mathematical variables. The free will of the experimenters and their random choice do not help here. Exactly one particular setting can be chosen for any particular space-time variable, while in Bell's derivations the same λ appears with all different setting pairs. It is frequently emphasized that the particle pair is emitted from the source before the setting is chosen and therefore is "independent" of the settings. The word "independent" is used here because the properties of the particle emitted at the source do certainly not change when the settings are switched. However, mathematical independence of two variables means that one can choose one variable arbitrarily from its domain (think of the domain as an urn containing the variables) and pair it with any element of the domain (a second urn) of the second variable. Such arbitrary pairing is, in general, not possible if the λs and measurement equipment exhibit some dynamics, some space-time dependencies.

It also is important for Bell's reasoning that he did not include any possible explicit time dependence of the measurement machinery. All that influences the experimental outcomes is the polarizer settings and the λ sent out from the source. Thus, Bell permits possible outcomes of these random variables such as $A_{\mathbf{a}}(\lambda_1) = +1$, $A_{\mathbf{c}}(\lambda_8) = -1$. The machines that Walter and I proposed are in addition exhibiting some space-time dependence that may be totally independent of the information that is sent out from the source. We, therefore, deal with a greater variability of random variables denoted, for example, by the symbol $A_{\mathbf{a}}^{st_n}(\lambda)$, as explained previously. In other words, we "sample" a larger space of likely (or unlikely) occurrences than Bell does.

In simpler language, Walter and I introduced "coins" or "machines" that are of different mathematical and physical nature. Our machines can, for example, be time-ordered and numbered by the measurement times, while Bell's machine outcomes are determined by random λs. Imagine a physical process that results in sending out white marbles for some period of time. Then, Bell's λ needs to be all the same during that time, because λ characterizes what was sent out. The measurement equipment, however, can at least in principle lead to different evaluations of the same object at different times and, therefore, lead to different correlations on the two sides. Such and other cases can be, for example, described by the mathematics of so-called stochastic processes (see below).

7.3 Probability

Walter knew, of course, everything about the foundations of probability, while I had only a very cursory understanding. That left for Walter a lot to explain. Interestingly, one cannot find much about the foundations of probability in the standard textbooks, and as mentioned above, some books just state that the word "probability" is easy to understand as long as one does not attempt to give a mathematically rigorous definition. But without mathematics, one cannot prove an inequality involving variables. For this reason, I give below a rather elaborate description of the important aspects related to the concept of probability. My friend Hans De Raedt commented that I might lose a lot of readers here because "it is just difficult stuff." Therefore, I have added the following summary of what is really essential to understand the remainder of the book. The detailed treatments of Kolmogorov's probability (often called classical probability) and of quantum probability can then be skipped by those less interested in the foundations of probability theory.

7.3.1 *Probability Essentials for EPRB Experiments*

The concept of probability is man-made and cannot be derived from sense impressions, just as the system of natural numbers cannot be derived from any sense impressions or data. To be sure, the natural numbers are closely linked to sense impressions and are derived from the procedure of counting. Probability is also linked to sense impressions, not as closely, though, as the natural numbers are. This is very important to remember for the discussions of Section 13.4.

Common language expressions for "probability" are the words "likelihood" or "chance." Some event in an "experiment," for example the tossing of a coin, is happening with a certain likelihood. A "fair coin" falls on heads or tails with equal likelihood. One says that the likelihood of the coin falling on heads is 50%, or that the number that "measures" the probability, the so-called probability measure, equals 0.5.

A second important relationship of the probability concept to observations is obtained by the concept of "averaging." If we perform an experiment over and over, N times with N being a very large number, and if we then add all results and divide by N, we obtain an average. For example, if we toss a fair coin 1000 times we might obtain heads 543 times and tails 457 times, or in another sequence of 1000 tosses we might obtain 491 heads and 509 tails. The average of the number of heads in the first case is 0.543 and in the second 0.491, and one can show mathematically that under certain, pretty general, conditions we will always obtain something close to 0.5 for a fair coin, which is equal to the probability measure that characterizes the outcome of the tosses.

The probability measure is thus a quantity related closely to observations. If we are sure that an event happens the probability measure equals 1, if an event is impossible it equals 0. Of course, there is nothing really sure and nothing really impossible in most practical cases. When we say the moon will be there tomorrow with probability 1, or probability measure 1, we do not really know whether a huge nuclear explosion caused by aliens will pulverize the moon, but ordinarily the moon will be there tomorrow. To work with probabilities in a rigorous mathematical way, we therefore do need to know or guess the possible outcomes. A very important concept related to this fact is that of the so-called sample space. The sample space is the set of all possible outcomes for all possible configurations of the machinery that we wish to describe with probabilistic methods. For a coin the possible outcomes are just heads or tails, and for roulette the possible outcomes are the 37 numbers of the roulette. The probability measure for the occurrence of any particular roulette number is $\frac{1}{37}$. The sample space thus links sense impressions, such as settings of the measurement machinery, and measurement outcomes, such as heads or tails, to the probability measure.

Quantum theory does not make use of a sample space. The concept of probability in quantum theory is mainly based on its predictions of the long-term average of measurements. A quantum particle is "prepared" by a certain experimental procedure and then a measurement is performed that detects the particle. If it is detected in EPRB experiments, we assign to the result the number $+1$; if it is not, we assign the number -1. As seen above, we can add all the results and determine the average of a large number of experiments.

That average is the main result that quantum theory predicts. The mathematics of the predictions of quantum theory is, therefore, quite different from the mathematics of conventional probability related to sample spaces. The experts know that quantum theory deals with so-called Hilbert spaces, which, in principle, have nothing to do with space-time coordinates. The quantum theory of Heisenberg and Bohr does not bother with predicting or defining probability measures for the single events; it only cares about the long-term average, while the mathematical connections of the single events have a lower rank of importance. This gives quantum theory a certain advantage, but also a certain lack of detail. EPR also pointed to complications. These complications are directly connected to the significant differences of the concepts of likelihood on one side and on long-term averaging on the other. This is best explained by the following "artificial" EPRB experiment.

In Tenerife, a magnetic coin is tossed in the field of a hidden magnet that is controlled by a secret clock and our function or app $rm(t_i)$ that we introduced in Section 4.1. This hidden magnet causes the coin to fall on heads for certain random numbers i of the experiments. For example, it may fall on heads for $i = 1, 4, 5, 6, 9, \ldots$, while it falls on tails for $i = 2, 3, 7, 8, \ldots$. Because the function is precisely random, the experimenter, not knowing about the magnet, will pronounce the coin to be fair as he measures averages of about 0.5 for heads and tails.

One can also perform such an experiment on La Palma and assume that the clock and computer driving the hidden magnet on La Palma have been synchronized at some time in the past with that of Tenerife. Both clocks and computers are very precise, just as atomic clocks are. The app in La Palma is assumed to be identical to that in Tenerife and is, therefore, representing the function $rm(t_i)$ also. A second experimenter on La Palma, tossing magnetic coins, will thus also conclude that the coin is fair. Now, however, a third experimenter introduces a time window for the measurements on Tenerife and La Palma and looks repeatedly at the outcome of one coin-toss on each island during that time window. For smallest time windows, the correlation is complete and the two coins on the two islands fall on the same side, either on heads or on tails. The time window, which is necessary for EPRB experiments, thus highlights the differences between likeliness of single events and averages over many measurements, the main prediction of quantum theory.

As an aside, this example also highlights another feature of EPRB-type experiments. Correlations of events on two islands can be explained in a variety of ways. In the above example, it does actually not matter whether or not the coins come from a common source to the islands. They do not need to have been entangled by originating from a common source. In contrast to this situation, we could generate somewhere two fast and parallel spinning coins and send one to Tenerife and the other to La Palma. They would arrive there still showing the same orientation like a gyroscope would and thus show perfect correlation when measured by certain equipment. Correlations can also be achieved by an infinity of examples that are in between the two just explained. Naturally, instantaneous influences at a distance and even spook could, at least in principle, also be at work to achieve the correlations.

Let me summarize the main points of this section by two remarks. First, all probability concepts are man-made and involve mathematics such as the averaging procedure. More generally speaking, if one is interested in the chance or likelihood of single events, one needs to know the sample space and the probability measures that may differ from the long-term averages. Second, time plays a very distinct role in probability theory, as we saw from the effect of the time window on the observation of correlations.

7.3.2 *Kolmogorov's Probability*

Kolmogorov developed a framework of probability theory by giving the ideas just discussed mathematical rigor based on a few axioms or "self-evident" truth. One cannot describe his theory in its full power in elementary terms, but it is possible to describe the portions of Kolmogorov's work that are relevant for the mathematical treatment of EPRB experiments.

Any strict mathematical framework needs to follow from a few axioms or rules about mathematical abstractions that describe events of nature. Kolmogorov used the axioms of the theory of sets as the basis for his framework. Many concepts of set theory are broadly known to everyone and even taught in elementary school. How can one connect such abstract concepts to actual experiments? This is achieved by the introduction of the sample space, the set of all possible outcomes: heads or tails for coin tosses or the 37 numbers of the roulette. These elements or objects of the sample space are denoted commonly by the Greek letter ω. For example, in the case of the roulette we have 37 possible outcomes and each of these outcomes is denoted by a specific ω, for example $\omega_0, \omega_1, \omega_2, \ldots, \omega_{36}$. These 37 objects are called the elementary events, because they correspond to the event of the roulette ball falling into a particular groove. We can imagine that Tyche, the Greek goddess of fortune (or Fortuna in Latin), will choose one of these elements randomly (for a fair roulette with equal likelihood) out of an urn and thus precipitate an actual event or outcome that corresponds to the dropping of the ball of the roulette in the groove corresponding to that number. This random choice of Tyche corresponds to a probability measure of $\frac{1}{37}$ for each of the possible outcomes. The sample space and the assigned probability measures lead us to the *probability space*, which forms the basis of all the calculations related to probabilities of the events. An event is the occurrence of a particular outcome that can be recorded at a given space-time point.

We also can consider composite events such as obtaining either a 1 or an 8 at the roulette table that happen at different points in time. How can we find the likeliness for such a composite event to happen? An important rule or axiom, necessary to answer this question, is the following. If we have two events such as ω_1 and ω_8 that are exclusive of each other and have no connection to each other (so-called disjoint events), then the probability measure for at least one of them to happen equals the sum of the probability measures of each separate event. This gives us the answer to the question what the likelihood is that we obtain either the 1 or the 8 in a trial with a fair roulette, which is $\frac{2}{37}$. If you have a fair coin, then the probability measure of obtaining heads is 0.5, while the probability measure to obtain either heads or tails is $0.5 + 0.5 = 1$.

To explore the full mathematical possibilities of dealing with such sets of events, Kolmogorov introduced *random variables*. Random variables are functions on the probability space such as $A(\omega)$. If Tyche choses a certain ω, for example ω_{34} from the "urn," that contains all the ωs, then we have $A(\omega_{34}) = 34$, indicating the mathematical result for the event. To describe the EPRB experiments by Kolmogorov's random variables, we need at least the 3 random variables $A_a(\omega)$, $A_b(\omega)$ and $A_c(\omega)$. If a certain ω is chosen by Tyche, say ω_{105}, then one obtains a result such as $A_a(\omega_{105}) = +1$, $A_b(\omega_{105}) = -1$, and $A_c(\omega_{105}) = +1$. Note that all functions on the probability space use the same ω. Why are the goddess of fortune and Kolmogorov using the same ω for all functions? The reason is that all these functions are defined on one probability space and, therefore, *by convention* Tyche chooses one ω for the given set of experiments that these random variables represent.

What is the advantage to use the random variables of Kolmogorov to describe the EPRB experiments or any experiments? At this point, there is no advantage, and Bell has in fact used intuitively similar functions with elements of reality λ instead of ω without any reference to Kolmogorov's probability theory; all that Bell used was physical intuition. We see, however, an important point here. If Bell introduces his algebra with elements of reality λ just by intuition, he needs to give a reason for using the same λ in all terms of his equation. If Kolmogorov uses the same ω, it is by convention because it is assumed that all the functions are defined on one probability space. Bell needs to show that his mathematical abstractions, the functions that describe the actual experiments, indeed form functions on a probability space that can be algebraically dealt with as we deal with numbers. This big step of invoking an algebra was assumed to be automatically true by Bell. How could he know that all events of EPRB measurements can be described by his functions of settings and λ?

In fact, Bell did not know. Key to that formation of a valid algebra from experimental events is that all the possible events can be subjected to logical "connections" such as "AND," "OR," and "NOT." One can then speak about a probability measure for the case that $A_a = +1$ AND $A_b = -1$ in a logical and consistent way (and the same must be true for the logical connections OR and NOT). In addition, one needs to make sure that a few other provisions are fulfilled. For example, if Ω denotes the complete sample space of all possible events, then for technical reasons we need to include also the so-called "empty set" as an event and element of Ω, as readers familiar with set theory will understand; the empty set of set theory is something like the 0 of a number system but not quite the same. One also needs to make sure that the probability measure of any event happening at all equals 1: if a coin is thrown it needs to fall either on heads or tails and the probability measure that any one of these two outcomes happens is 1. This may appear all a little complicated to the inexperienced reader. However, all we need to remember is that the events need to be very carefully examined to make sure that they can be treated in an algebraic fashion. Is there any way to figure out whether the random variables of complicated experiments form an algebra, even if we do not know all the details? This will be discussed in Chapter 13.1. Here we give just a little preview.

Consider doctors who examine patients under certain circumstances that are denoted by **a**, **b**, **c** ... and note their diagnosis as $A_{\mathbf{a}} = +1$ for a positive diagnosis under circumstance **a** on one island and $A_{\mathbf{c}} = -1$ for a negative diagnosis on another island. Can we say whether or not the As form an algebra without knowing more about the disease? Interestingly enough, the famous mathematician Boole asked himself this question in 1862, as I found out during my collaboration with Hans De Raedt and Kristel Michielsen. Boole came up with an answer. If the functions A fulfill what we now call Bell's inequality, then the use of the functions A in an algebraic way is "kosher." If this inequality is violated, something is wrong and we need to know more about the disease than just the circumstance **a**, **b**, **c** .... Of course, Boole knew nothing about quantum physics and what inequality Bell would come up with. However, he showed clearly that an almost identical inequality should be used to figure out whether one can use an algebra for random variables like our functions A. Bell worked the other way around. He used the algebra just intuitively to start with and derived the inequality for the random variables that he claimed must then be valid within his assumptions.

We can see from the above that the mathematical definition of the word probability has many components. We need to have a well-defined sample space of possible outcomes (such as the numbers of the roulette), we need to be able to deal with the outcomes in a logical way (apply logical operations AND, OR, NOT), and we need to be sure that the events can be dealt with algebraically. We also need to be able to assign a probability measure to all these outcomes. To derive this measure, we need to know *all* the details of the experiments that determine all the possible outcomes.

It is not surprising then that the word probability is impossible to precisely define without such mathematical details. Speaking from the point of view of both mathematics and everyday language, we wish to extract from probability considerations two "assurances." First, we wish to know the approximate long-term average, the result of large numbers of experiments over time. In the case of the roulette, with each number having a probability measure of $\frac{1}{37}$, we will obtain any particular number in N experiments (where N is very large, e.g., 1 million) about $\frac{N}{37}$ times. The second point we like to know is how likely any single number will be hit for any single event, and that likelihood in case of a fair roulette is also 1 in 37 or $\frac{1}{37}$, which is the probability measure. Thus, for a "fair" roulette, the two assessments or predictions are very closely related. For an "unfair" roulette, however, the two assessments mean something very different. Assume, for example, that a crooked Casino-owner has magnets that guide the roulette ball to hit certain numbers. By suitable design this owner can make sure that on the long run all numbers come an average of $\frac{N}{37}$ times. However, for a single event he can choose a certain number with probability 1 instead of a likelihood of $\frac{1}{37}$, because the magnet guides the ball to that number in that moment. Applying this knowledge to the EPR experiments, we see how difficult it is to define a probability space and probability measure if we do not know all the factors that influence the experiments. This problem is also very well known in the medical community that deals with the statistics and probability of diseases.

If, on the other hand, Bell's functions $A_{\mathbf{a}} = \pm 1$ are assumed to be functions on a Kolmogorov probability space (random variables), then one can deal with them by using the algebra that applies for integer numbers and faithfully describe the experiments with all their logical implications. Bell's inequality appears thus in a new light. If we replace Bell's λ by abstract mathematical elements ω of a probability space, then we obtain

$$A_\mathbf{a}(\omega)A_\mathbf{b}(\omega) + A_\mathbf{a}(\omega)A_\mathbf{c}(\omega) - A_\mathbf{b}(\omega)A_\mathbf{c}(\omega) \leq +1. \qquad (7.1)$$

The fact that the same ω is used in all terms now follows automatically, because all functions are defined on the same probability space. Thus, Bell's inequality is automatically true if the functions A are random variables.

Equivalently if the experimental outcomes can be sorted in lines such as shown in Eq. (7.1) above with the same ω or λ in every term, then Bell's inequality is valid also. Because Bell assumed that possibility of sorting from the start, his inequality was automatically valid. There was no need for him to assume Einstein locality or other physical conditions to derive the inequality. Bell's basic assumption that λ and the settings $\mathbf{a}, \mathbf{b}, \mathbf{c} \ldots$ are independent mathematical variables is already sufficient to derive the possible sorting of Eq. (7.1) and the validity of the inequality. However, as explained previously, λ and the settings cannot be considered independent mathematical variables if general elements of reality that include space-time are involved. Remember that this is the reason why we write the settings as subscripts.

It is interesting that the mathematical elements ω must, according to Kolmogorov's probability theory, depend on the settings, because the sample space depends on the settings. Bell's inequality is still valid as long as the ω are elements of the sample space and as long as the functions A are defined on the probability space related to that sample space. Remember that Bell assumed that his λ does not depend on the settings. Bell thought that this requirement was necessary to derive his inequality. As we know, Bell wished to guarantee with this requirement Einstein locality, because this way λ cannot depend on the settings of the other town. As we see now, that requirement is not needed anymore to prove the inequality as long as we deal with functions on one probability space.

It is true, however, that influences at a distance may invalidate Bell's inequality. This fact can be formulated in a Kolmogorov type of notation by using different functions. The function $A_\mathbf{a}(\omega)$ on Tenerife must now be denoted by $A_{(\mathbf{a},\mathbf{b})}(\omega)$ because the setting $\mathbf{b}$ on La Palma exerts an influence. $A_\mathbf{a}(\omega)$ in Tenerife becomes $A_{(\mathbf{a},\mathbf{c})}(\omega)$ if the setting on La Palma is $\mathbf{c}$. Invoking similar arguments for all the settings involved, we obtain instead of Eq. (7.1) the following equation:

$$A_{(\mathbf{a},\mathbf{b})}(\omega)A_{\mathbf{b}}(\omega)+$$
$$A_{(\mathbf{a},\mathbf{c})}(\omega)A_{\mathbf{c}}(\omega) - A_{(\mathbf{b},\mathbf{c})}(\omega)A_{\mathbf{c}}(\omega) \leq +3. \tag{7.2}$$

The left-hand side is now only smaller or equal to 3 because all the functions can be different unlike the functions in the Bell inequality. This equation is always valid and puts no limitations on the experiments.

However, the violation of Bell's inequality necessitates by no means instantaneous influences at a distance. Space-time elements of reality such as the measurement times are not included in the conditions that validate Bell's inequality. As we know, time has a certain ordering and cannot necessarily be regarded as a random variable. Space-time also does not only describe static colored or flavored marbles, but describes the dynamics and the relation of colored and flavored marbles, of elementary particles, to each other. The general importance of the dynamics was also emphasized by the mathematician Accardi. The probability theory of Kolmogorov, and every probability theory that describes nature, needs, therefore, to treat time in a special way that also includes the "ordering" of events in a timelike fashion. As mentioned, Kolmogorov's way is the use of stochastic processes with functions A that depend on time or more generally on space-time.

Reading a number of standard probability texts I learned that a stochastic process is described by random variables whose statistical properties change with time. This device of probability theory was found necessary to describe the dynamics of physical processes. Dynamic processes were exactly what I had in mind when introducing time dependencies of the measurement instruments and photon sources of EPR experiments. In Section 3.3.2 we introduced such a time dependence by using functions such as $A_{\mathbf{a}}^{st_i}(\lambda)$ for the measurements taken in one wing of the EPR experiments, say in Tenerife. I asked Walter about our change from time to space-time: "Can one replace the time in standard stochastic processes by space-time?" Walter explained that as long as one considered the measurements as countable, there was no problem. Because we use $i = 1, 2, 3, 4 \ldots$ in our notation, we have assumed that we could number the measurements by natural numbers and, therefore, they were countable. One could thus just use the numbers *i* instead of time or space-time. Remember that, because our random variables depend also on the settings, we have used space-time as an index to avoid the pitfalls to regard time, space-time and settings as independent mathematical variables. Can we include also the fact that we have two distant experiments and still have a stochastic process? Walter assured me that this was no problem either. "Vector stochastic processes have been often considered, which means just that we can have a pair of outcomes described by the pair of functions such as $A_{\mathbf{a}}^{st_i}(\lambda_i) A_{\mathbf{b}}^{st'_i}(\lambda_i)$. This pair describes, then, a dynamic physical process in two measurement stations by use of random variables. It became clear to me that the definition of a stochastic process was so general that almost any time-dependent process that involved probability could be described by it, not only special cases such as Brownian motion. The result of the model that I discussed in Section 4.2 can be restated using the concept of stochastic processes as follows: for every sequence of measurements with randomly chosen setting pairs (**a**, **b**); (**a**, **c**) and (**b**, **c**), one can find at least one vector stochastic process that results in a long-term measurement average that agrees with the predictions of quantum mechanics.

7.3.3 *Quantum Probability and Why Einstein Opposed It?*

The probability of any event, any macroscopic occurrence that can be recorded by man or machine as data, can be represented by the framework of Kolmogorov if all the possible "workings" of the event-producing "machinery" are known. Then, the likelihood of an event as given by a probability measure can be assessed. What is the difference between Kolmogorov's probability and quantum probability? The answer to this question was originally given by some of the fathers of quantum mechanics, particularly by Born and Bohr, and was later presented within a rigorous mathematical framework by the great John von Neumann. We use the acronym BBN for this framework. Recently a more modern answer has been developed by Gell-Mann, Griffiths, Hartle, Omnes, and others that is called the consistent history approach, for which we use the acronym GGHO.

The difference of the BBN version of quantum probability from Kolmogorov's is at least twofold. For one, classical objects that are being measured are well-defined macroscopic entities and can be endowed with space-time coordinates. Examples are the numbers of a fair roulette, or head and tail of a fair coin. The objects of the quantum world are often idealized as elementary particles with properties such as "color" and "flavor" that remind us of our macroscopic world. A characterization by space-time coordinates, however, must account for the Uncertainty Principle. This fact presents a difficulty for the dynamic description of quantum particles. The difficulty is resolved by defining and clearly separating compatible measurements (linked to so-called commuting operators of the mathematical theory) and mutually exclusive incompatible measurements (linked to so-called noncommuting operators). In addition, a preparation procedure is introduced. This procedure precedes the actual measurement and defines the so-called quantum state of the particle that is mathematically described by the wave function. In the case of the EPRB experiment the particles are entangled photons. The preparation is accomplished by the excitation of electrons of certain atoms to a higher energy state from which they cascade down in two steps emitting the correlated photon pairs. Then the measurement is performed with the help of polarizing equipment oriented in certain directions. Thus, measurement and preparation are both necessary for a well-defined quantum experiment.

Second, quantum theory in the BBN interpretation does, in general, not deal with single events but only provides rules to compute averages over large numbers of measurements. These rules do not introduce or use the algebra discussed above that involves logical operations for the single events. In fact, quantum theory often "forbids" logical dealings with the single events. It is true that the researchers of quantum mechanics have spoken on occasions about the likelihood of single events that they deduced from the quantum predictions for the long-term averages. However, this would only be permitted if we know that the machinery is "fair" and does not have any unknown hidden time or space-time dependencies. Remember the magnetized "unfair" coin. Someone not knowing about the magnetic trickery cannot tell the likelihood of any outcome of a single experiment just from knowing that on average we measure as many heads as tails.

Some experts may deny the existence of such hidden dependencies for the EPRB experiments. However, it is clear that the incoming photons "shake" the electrons of the polarizers and cause a time-dependent response of these electrons according to both classical and quantum mechanics. The interaction of photon and equipment electrons will occur during a time period that must be dependent on the spatial arrangements of the polarizers leading to time and setting dependencies for the transmission, reflection, or absorption of the photons by the polarizers. Furthermore the quantum particles and the measurement equipment interact with the environment such as the rotating earth. Experts are encouraged to consider in addition the dependence of quantum experiments on the vector potential and to think of the fiber bundles of modern gauge theories. Thus, there is no reasonable argument that can deny that the dynamics of the measurements may have to be characterized by more than just the spatial direction of the polarizers that is indicated by the vectors **a**, **b** and **c**.

Therefore, it is clear that the quantum mechanics in the BBN interpretation does not provide, and does not intend to provide, an algebra for the single events. It provides only an algebra to determine the long-term averages. How does it calculate the long-term averages? To do this, quantum mechanics uses two constructs. One is that of the quantum state, experimentally defined by the preparation procedure and mathematically represented by a wave function ψ, a complex number that is a function of space-time $\psi = \psi(x, y, z, t)$. ψ develops in space and time according to certain rules (the Schrödinger or Dirac equation) until a measurement is taken. That measurement is characterized by mathematical operations that are determined by the measurement settings such as **a**, **b**, **c** . . . for the polarizers and by the particle property (such as the spin) that is measured. One describes these mathematical operations by so-called operators that are denoted by symbols such as σ_a. The operators act on the wave function that depends on the space and time variables. This mathematical separation of settings (related to the operators) and space-time (related to the wave function) is in clear contrast to the combined use of settings and λ as the independent variables of Bell's functions $A(\mathbf{a}, \lambda)$, $A(\mathbf{b}, \lambda)$, $A(\mathbf{c}, \lambda)$.

Thus quantum mechanics separates the space-time description from the measurement settings and does not assume any independence of settings and space-time variables in any functions. The operators characterized by the settings cannot even be applied in arbitrary order to the wave function. One says that they do not commute. The fact that the operators do not commute or cannot be interchanged in a sequence of experiments adds a certain order to the events that is important in the consistent history approach of GGHO.

The connection of quantum theory and long-term averages is, according to Born, the following. We consider only the measurement of the location of a particle at a certain point in space. This corresponds to describing the measurement procedure with a mathematical operator (or operation) corresponding to the space coordinate. We do not describe this operation any further. It acts on the wave function ψ according to rules that quantum theory provides. The whole procedure of such space-location measurements and calculations proceeds then as follows. First, one prepares a quantum entity (particle) so that it is described by a wave function $\psi(x, y, z, t_0)$ at time t_0. The wave function develops then until the point of measurement at time t_1 to equal $\psi(x, y, z, t_1)$. Second, one performs a measurement at that position x, y, z, t_1 and determines whether one finds a particle there. The chance C for finding the particle is computed from the value of the absolute square of the wave function at the point of measurement, which is $C = |\psi(x, y, z, t_1)|^2$. If you repeat that process identically N times, where N is a very large number, you will find the particle about $N \cdot C$ times. This means, as we know from the example of the roulette, that C takes the role of a "probability measure" in the sense that you can compute with it long-term averages.

We can see here that quantum mechanical measurements and theoretical descriptions of these measurements differ in two major ways from what we do ordinarily when measuring macroscopic objects. First, we have to *prepare* a particle in certain ways that let us determine its wave function. We cannot say that we have a marble at certain space-time coordinates but just deal with a quantum entity prepared in such a way that we know the mathematically abstract wave function. Second, we perform a measurement but we cannot say anything about the single outcome; we have only a recipe for long-term averages. This gives us the following complication that we do not have in our ordinary macroscopic world and measurement procedures. We cannot easily perform and evaluate any sequence of measurements, because that would leave out the preparation step and we would not know with what wave function we are dealing during each step of the sequence. It is important to realize that, therefore, it is difficult to include a time-ordered process such as a stochastic process with a sequence of measurements and measurement outcomes into the framework of the Copenhagen school.

The modern consistent history approach accomplishes just that (Griffiths, 2002). They assign probabilities to whole time-ordered histories. Their procedure is beyond what can be explained in lay terms, because quantum theory uses a space different to the space that surrounds us, the so-called Hilbert space. The Hilbert space, named after the famous mathematician, is a vector space very similar to the vector spaces we remember from high school, systems of objects that have a direction and a length just as an arrow. Einstein's four-dimensional space-time also forms a vector space. The Hilbert space, however, can have any number of dimensions. Its vectors correspond to the states of quantum mechanics. The wave function discussed above represents such a state and can also be mathematically represented by vectors in Hilbert space. The question is now, how can these vectors of Hilbert space be related to things that happen in our space-time and that we measure with our yard-sticks and clocks? Or, from the view of probability theory, how do histories of wave functions relate to the events in space-time that may be described by a Kolmogorov-type stochastic process?

This is what Robert Griffiths discusses in his book *Consistent Quantum Theory* (Griffiths, 2002) and reaches beyond the elementary discussions presented here. However, only a small subsection of the GGHO framework is necessary to answer questions about whether Einstein was right and, in my opinion, no part of it is necessary to understand my objections to Bell's work. It is reassuring, however, that Griffiths explicitly refutes the necessity of actions or influences at a distance. The portion of the GGHO approach that is relevant to EPR will be discussed later together with other modern views of the quantum masters. Here we finish this section by explaining the important reason why it is difficult to connect any histories of certain wave functions to Kolmogorov probability.

The reason is that quantum mechanics does not let you define a sample space in a straightforward way. The basic problem can be understood from Einstein's example of the ball in the two closed containers, with a quantum wave function that describes the fact that the ball is either in the one or in the other container. Such a wave function is called a superposition of the quantum states. Quantum mechanics must have such superpositions, because one must be able to add or subtract the states and therefore the vectors of Hilbert space. No vector space can be defined or is of any use without the possibility of adding and subtracting vectors.

The two photons in the EPRB experiment are exactly in such a superposition of Hilbert-space states that are nowadays called "Bell states." Photons that are generated may be horizontally (H) or vertically (V) polarized. If we give an index 1 to one photon and 2 to the other of a photon pair, then we have the 4 possibilities that photon 1 and 2 are vertical (V_1, V_2), photon 1 is vertical and 2 horizontal (V_1, H_2) as well as the two more possibilities (H_1, V_2) and (H_1, H_2). This would be all that we need to know for a description of the vertically and horizontally polarized photons, if we describe them as we are used to describe things of our macroscopic world. Quantum theory, however, describes states of the photons by vectors in Hilbert space. Such a vector may, for example, be written using an arrow type symbol such as $|V_1\rangle$, meaning the state of particle 1 corresponds to a vertical polarization. Or we may write $|H_1, V_2\rangle$, meaning that the state of particle 1 is horizontal and that of particle 2 vertical.

As mentioned, a cornerstone of quantum mechanics is now that the vectors of Hilbert space can be added or subtracted. This leads us back to the so-called Bell states:

$$\text{Bell state} = |H_1, V_2\rangle - |V_1, H_2\rangle. \qquad (7.3)$$

This state represents a so-called superposition and means that if the particle 1 is horizontal, then particle 2 is vertical. If, however, particle

1 is vertical, then 2 is horizontal. The state does not give away which particle has which polarization. It only describes what polarization of one particle will be measured, conditional to what polarization was measured for the other particle. The same is true for another Bell state that we have actually talked about all along starting in Section 1.2:

$$\text{Bell state} = |H_1, H_2\rangle - |V_1, V_2\rangle, \qquad (7.4)$$

with the only difference that now we measure on both sides the same type, either H or V, but we do not know in advance which it will be. The experienced reader may have seen these two latter equations with a factor of $\frac{1}{\sqrt{2}}$ in front. This factor is a so-called normalization that one needs for the precise algebra of long-term averages.

For a mind not trained in quantum theory such superposition states are very strange "states," while for the quantum expert they are entirely natural. Quantum mechanics tells us that if we can prepare such a state and then perform a measurement, then the outcome on one side is entirely random. This means that if we position the polarizer for particle 1 (say that in Tenerife) to horizontal, then the particle may or may not propagate through the polarizer to the detector and the result is entirely random. However, if the detector clicks and thus indicates that particle 1 was horizontally polarized, then we know for sure (with probability 1) that a horizontal polarizer in La Palma will also lead to a click, presuming that the wave function was given by Eq. (7.4). The miracle is, of course, that indeed such a Bell state can be prepared, and this is exactly what the Aspect and Zeilinger groups did. They confirmed by their measurements that the outcome on each side is very random, while the outcomes on the two sides are strongly correlated.

Einstein added to this situation the following conundrum that we describe in a modernized variation. Let's say that the two measurement outcomes of Tenerifa and La Palma are fed into a computer that transmits a signal to Paris if and only if both detectors click. In Paris we have a plastic explosive that can be triggered to go off by a receiver that reacts to the computer signal from the Canary Islands. We denote this plastic explosive by the symbol P_{pl} as long as it has not exploded and by P_{ex} after it has exploded and forms a cloud over Paris. Indeed, as we know, such explosions happen and a plastic explosive may still be plastic at one time t_1 and a hot cloud of smoke at a later time t_2. However, if we assume that quantum theory is completely valid also for macroscopic objects such as the plastic explosive, then the plastic explosive exists in a superposition of states of exploded and unexploded plastic, corresponding to the Bell state and the two measurements in Tenerife and La Palma. The reader may be more familiar with the superposition of a dead and an alive cat that Schrödinger had described after discussions with Einstein.

Einstein maintained that no reasonable theory could describe the physical reality in such a way. He maintained that there must be an element of physical reality connected to the particles and measurements that leads to the correlations and to the actual measurement outcomes and that quantum theory does not tell us about it and is, therefore, incomplete. The superposition is, thus, according to Einstein, only a mathematical construct that helps us calculate the outcomes in a very efficient way and without involving space-time but using probability concepts instead. The modern GGHO interpretation of quantum theory also regards superposition as a construct related to a "pre-probability." It is, fortunately, not necessary to understand all these details in order to understand my objections against the validity of Bell's work. It is, however, important to remember that many of the quantum experts deny the existence of elements of reality, often called hidden variables, on the basis of Bell's work and not on the basis of the framework of quantum theory itself.

Chapter 8

New Friends, New Inequalities

Allein sie haben schrecklich viel gelesen.
Wie machen wir's, dass alles frisch und neu
und mit Bedeutung auch gefaellig sei?

—Goethe, *Faust* I

Free translation:

Alas, they read an awful lot.
How can we put it fresh and hot,
so that it pleases and projects
the great importance of the facts?

Armed with this more detailed understanding of the basic probability concepts and their relation to EPRB experiments and Bell's work, I was ready to attend the conferences for which we had received invitations. The letters of invitation were also very reassuring and showed that there was quite a big community that had significant issues with Bell's work.

Einstein Was Right!
Karl Hess
Copyright © 2015 Pan Stanford Publishing Pte. Ltd.
ISBN 978-981-4463-69-0 (Hardcover), 978-981-4463-70-6 (eBook)
www.panstanford.com

8.1 Växjö

Andrei Khrennikov was born in Russia and holds a professorship in mathematics in Växjö, Sweden. There he organized a number of conferences on the foundations of probability and quantum physics (see, e.g., (Khrennikov, 2002)) and wrote and edited several books and book chapters on the same topic (see, e.g., (Khrennikov, 1997)). It was he who had invited us to the Växjö conference, and when Walter and I arrived he greeted us at the airport. Walter and I liked Andrei from the first moment on and we soon became friends. Andrei was undoubtedly the mathematician with the greatest expertise in the area of problems related to the foundations of probability and quantum theory, and particularly for EPRB experiments. He had close interactions with Luigi Accardi, who was another mathematician with detailed experience in the EPR area. I have previously mentioned Accardi's chameleon model. He lets Bell's λ change like a chameleon once it hits the polarizers (or any other measurement equipment). The dynamic and adaptive behavior of λ is the key to his mathematical theory of violations of the Bell inequality. Accardi had a bet with Richard Gill and hoped that he could develop a mathematical computer program that could play the Bell game. Such a program was never fully demonstrated as far as I know, and I suspect that there was a problem to achieve equal outcomes for equal settings on both sides when using the chameleon effect. Simply said, if you have dynamically changing chameleons on each side that do not know from each other, then it is very difficult to guarantee that for equal settings they have equal color. Nevertheless, Walter and I had very interesting discussions with Accardi. He was considering the dynamics of EPR experiments in a mathematically abstract and rigorous way, and thus time played a major role in his work. This fact reassured me about my conviction that a space-time dependency of the measurement equipment was the only way toward Einstein local violations of Bell's inequality.

It was a great joy to meet with these scientists who were completely open to the idea that Einstein was right. Our hotel and the conference facilities were located around a fairly big lake and we often walked for hours around this lake discussing Bell's work. Several very well known scientists made presentations at Khrennikov's conferences, including L. E. Ballentine, L. Hardy, B. Hiley, N. D. Mermin, I. Pitovsky, and A. Plotnitsky, and we found that the opinion on the work of Bell was not as unevenly split as it had appeared to us after our *PNAS* publication. There was, however, a separation of the participants with opposite opinions, similar to that of water and oil, in spite of Andrei Khrennikov's great efforts to stimulate togetherness. I never saw Hardy in lectures of the anti-Bell group, and I think that he and others counted everyone who opposed Bell as, at least slightly, crazy. More open-minded participants, as for example Ballentine, did attend the Bell-opposing lectures, but did not comment. I tried to involve Ballentine in a discussion during lunch. However, he just looked at me and let me know that Bell's work was a serious topic that could not be discussed during a short lunch break. Nevertheless, Walter and I felt invigorated by what we heard and left with the conviction that we should continue on our path and fight.

8.2 Jackson Hole

The conference in Jackson Hole, Wyoming, was organized by Marlan Scully , who was one of the referees of our *PNAS* paper and therefore knew about our work. This conference was on the specific topic of quantum optics including EPRB experiments. The attendants were all optics experts and Walter and I benefited greatly from discussions and from learning about experimental details. I realized for the first time that the experiments of the EPRB type were, regarding their precision and statistical error, very different from the experiments that were the pride of quantum optics. The energies of the optical spectra, of the light emitted from atoms, can be determined to more than 6 digits. Some quantum properties can be reproducibly measured to 13 digits with accuracy still increasing. Quantum mechanical theory explains and predicts these very precise measurements. The correlations of the EPRB experiments are of a totally different kind. The predictions of quantum theory and the experiments of Aspect, Zeilinger, and others differ overall by 10% to 15% or more. From the discussions in Jackson Hole, I got the conviction that these were not just experimental uncertainties that could be improved. There is something more basic involved here. I would return to this point when discussing the work with Hans De Raedt and Kristel Michielsen.

My work with Walter and our presentation in Jackson Hole was well received, and Marlan Scully was very laudatory about it and said something to the effect that our presentation was amazing. Sylvia was standing right next to me at this moment and said, "Of course." Walter just laughed and said, "Unfortunately, not everyone thinks so." The conference was also attended by an editor of the prestigious journal *Physical Review Letters*. He was leading an interesting discussion on why it was so difficult to publish new views of scientific topics, and pointed to the problem of choosing unbiased referees. Experts are often totally convinced of the standard ways and that every new view is just bogus, and correspondingly they write very destructive reviews. What should a journal editor, with just a cursory understanding of the subject, then do? No one had any solution. I had known about this problem from all my previous publications in a variety of areas. However, both Walter and I saw now that things were even worse in the area of the foundations of physics. Walter was surprised about the very strong reactions of some of the "mandarins" of the foundations of physics to anything contradicting their views. In Växjö, we got the impression that many researchers of this area were less inclined to appreciate the views of others.

One could say, but this may be unjust, that Bell set the stage in his paper "How to Teach Special Relativity," in which he included virtually no reference to Einstein. Bell also never noted that major facts related to his inequality had already been considered before him in great detail, by the mathematician Vorob'ev in 1962, by George Boole hundred years earlier in 1862, and by others. The author is well aware that one can easily miss a paper, because of the enormous number of publications in any area. It is, however, another matter if one does not find relevant references during a time span of more than 25 years. Walter said sarcastically, "Maybe they all want to get the Nobel prize."

Jackson Hole is a very beautiful place embedded in a sensational mountain world. Here, Sylvia and I saw a different side of Walter:

Walter the mountaineer. We made frequent excursions to the top of some of the mountains by using chairlifts and cable cars. Actually it was mostly Sylvia and I who used the chairlift. Walter just looked at us and told us that the lift was too expensive and he would hike up the mountain. I looked up the steep incline, interrupted by some vertical rock walls, and just bought the tickets for the chairlift. "We will wait for you up in the restaurant," I said to Walter, and on the chair I said to Sylvia, "He will never make it." We walked around on the top of the mountain to some easily reachable lookouts and then went to the restaurant. It did not take very long, and Walter was also there. He had climbed very fast. When we took the cable car to a much greater height, Walter took the ride with us up the mountain, but insisted that he would hike down. This time he beat us and was down before we arrived and waved to us from the balcony of the restaurant that was located in the cable car station in the valley. I knew from a few remarks of Walter that he was a dedicated mountaineer in his youth. It surprised me that he still could perform like that at his present age. The next morning, however, Walter could barely get out of bed and had cramps in his calves. "Well," I said to Sylvia, "I am surprised that he can walk at all." I realized only a few years later that Walter was a well-known mountaineer. In 2006 he was invited by Austrian mountaineers to visit the Dolomites at the occasion of his 70th birthday and to visit a number of places where difficult rock-climbing routes were named after him. Walter never returned from that tour.

8.3 New Inequalities: Vorob'ev Cyclicities

Up to this point, we knew that there was Bell's inequality and at the conference in Växjö we also heard about a number of other Bell-type inequalities such as that of Clauser–Horne–Shimony and Holt (CHSH), which is discussed below. We did suspect that a lot of such inequalities could be formulated, and Walter thought that the mathematical literature must contain some mentioning of such inequalities. After talking to many colleagues at the just described conferences, he did a search of the literature to see if mathematicians had previously dealt with such problems. He found the astounding work of Vorob'ev (Vorob'ev, 1962), who had figured out all the possible ways of obtaining Bell-type inequalities about two years before Bell published his paper.

N. N. Vorob'ev's work demonstrated the fact that there exist correlations of three or more random variables for which no sample space and no probability measure and thus no probability space can be found that explains these correlations. The actual mathematical treatment of Vorob'ev, and that of some of his precursors on the topic that Walter also found, is mathematically too involved to be presented here, and I try to just report the essence of it. This essence, it turns out, was also found by Boole hundred years earlier. Walter did not find Boole's work in spite of the fact that the Austrian physicist Karl Swozil had pointed us toward it. We were so excited to learn about Vorob'ev's work that Walter stopped his literature search and we forgot about Boole for the time being. This was a mistake, because Boole's work contained several important revelations, as I learned later from Hans De Raedt. Another chapter is devoted to this topic.

What Vorob'ev reported is closely related to the algebra of events that we discussed in the Section 7.3.2. Without that algebra and functions of this event algebra, the random variables, probability theory does not have much power. We have seen that in order to obtain such an algebra, we need to be able to handle the events with the tools of logic and need to be able to subject them to the logical connections (AND, OR, NOT). Without this ability, one cannot expect to deal with the mathematical functions, the random variables, the way we deal with numbers and the Bell inequality does not make any sense.

The great contribution of Vorob'ev is that he traced the ability or inability to construct an algebra for the events to a general method involving random variables. That method relies on inequalities similar to Bell's. Vorob'ev found that expressions containing random variables and a "cyclicity" must fulfill certain inequalities. What is a cyclicity? Consider Bell's inequality as presented during Tony's seminar,

$$AB + AC - BC \leq +1 \qquad (8.1)$$

and note the following. If we insert values of ± 1 in the first two products, then we have chosen all the values of random variables appearing in the inequality. Thus the values of variables B and C in the third product are fixed and we cannot freely choose them anymore. This is what Vorob'ev called a cyclicity. Other examples of cyclicities are

$$AB + AC + DB - DC \leq +2, \tag{8.2}$$

and

$$BC + BD + AC - AD \leq +2. \tag{8.3}$$

If the values of the random variables are chosen for the first three products in Eqs. (8.2) and (8.3), then those of the fourth product are completely determined. By inserting all possible values for A, B, C, and D, you can convince yourself that the inequalities must always be fulfilled. These latter inequalities are called the CHSH inequalities. Using Vorob'ev's recipe, everyone familiar with elementary algebra can construct any number of such inequalities that need to be fulfilled for the random variables on a probability space. Vorob'ev generalized the procedure just described by introducing "topological-combinatorial cyclicities." The physicist Pitovsky presented similar but less general results (Pitovsky, 1989) independently.

One can summarize Vorob'ev's findings approximately in the following way. Given any sample space, probability measure, and random variables, an event algebra can be constructed and used if and only if the inequalities for all possible cyclicities are fulfilled on average for the measurement outcomes. In other words, a set of random variables such as A, B, C, and D can only then be defined on a probability space of Kolmogorov if we do not violate inequalities of the kind discussed by Vorob'ev and, of course, by Bell for the long-term measurement averages. For the particular case of EPRB experiments one can show that the possibility of defining four random variables A, B, C, and D on one probability space is equivalent to the fulfillment of all inequalities of the CHSH type averaged over many measurements. Bell's inequality is just a special case of CHSH. The work of Walter and myself that proves this fact with mathematical rigor (Hess and Philipp, 2006) is only published in German, because we had problems to publish our work in the English literature—a sad fact that I discuss further below.

If one assumes, therefore, that one deals with four random variables on one probability space, then one automatically assumes that the CHSH inequalities are fulfilled and with them the Bell inequality. In mathematical language this fact can be re-stated in the following way. Given a sample space Ω with elements ω, elements of the type of colored and flavored marbles in the urn with appropriate probability measures, and given four random variables $A(\omega)$, $B(\omega)$, $C(\omega)$, and $D(\omega)$, these random variables fulfill the CHSH and, therefore, the Bell-type inequalities.

Vorob'ev's findings show a clear inconsistency of Bell's approach. Bell requires that his λ not depend on the polarizer- or any equipment-settings. This is his way of making sure that Einstein's limitation to the velocity of light in vacuum or lower is obeyed. The fast switching of the settings, before the entangled pair arrives, guarantees then that no influence can arrive from the other wing of the experiment at the moment of measurement. However, why should λ not depend on the local settings, the settings that the particle interacts with? From Vorob'ev we know further that even if λ depends on all four polarizer settings $\mathbf{a}$, $\mathbf{b}$, $\mathbf{c}$, and $\mathbf{d}$, the Bell and CHSH inequalities are still valid as long as A, B, C, and D are functions on one probability space. Instantaneous influences at a distance do not matter as long as we can find one common Kolmogorov probability space for the functions A, B, C, and D. Only if the instantaneous influences lead to different probability space can the inequalities be violated. However, any physics that demands the use of more random variables or the use of more than one common probability space may lead to violations of Bell-type inequalities as well.

As discussed all along, the EPRB experiments may just not be describable by four random variables (functions) at all. We may need to use more functions and include a space-time index. Using these facts one can disprove any and all the proofs of Bell and his followers that can be found in the literature. Walter and I actually went through this exercise, and we published some of our refutations in (Hess and Philipp, 2004) and later more in (Hess and Philipp, 2006).

Chapter 9

Bell's Many Proofs

Was hilft's wenn ihr ein Ganzes dargebracht
Das Publikum wird es Euch doch zerpfluecken.

—Goethe, *Faust* I

What good is it to fashion and present a whole?
The audience will pick it all to pieces.

Building on Vorob'ev's work, we thought it should be easy to explain why Bell's reasoning failed and why one should look at Bell's "proof" from a different angle. However, we soon found out that the followers of Bell dismissed our counterarguments because there existed a large number of different proofs, most of them explained in Bell's collected works (Bell, 2001). As mentioned earlier, Walter and I, therefore, decided to go through all of these proofs and to attempt to refute each single one. Our work was performed over several years and the presentation here does not follow the precise timeline. We found that all of these proofs were actually based on a few principles that can be described and refuted rather easily, and here is an excerpt of our findings.

Einstein Was Right!
Karl Hess
Copyright © 2015 Pan Stanford Publishing Pte. Ltd.
ISBN 978-981-4463-69-0 (Hardcover), 978-981-4463-70-6 (eBook)
www.panstanford.com

9.1 Lille, Lyon, and Dr. Bertlmann's Socks

Bell presented his inequality in a more general way, using so-called conditional probabilities instead of just the outcomes of the functions A. While Bell added generality in this particular approach, he did not extricate himself from his conceptual net and still considered the functions A as functions of two independent mathematical variables such as $A(\mathbf{a}, \lambda)$, with each λ appearing about equally frequently with each setting $\mathbf{a}$, $\mathbf{b}$, and $\mathbf{c}$. Note that we reverted here to Bell's original notation with two independent variables that is, as reiterated already so often, incorrect if space-time is included. If settings and λ are assumed to be independent, the Bell inequality is valid to start with, no matter what other physics or mathematics is introduced. Bell's variations of his proof are presented here mainly for the reader interested in the details.

Bell considers the joint probability for obtaining two specific outcomes of the measurements at two locations. In Bell's notation we have, for example, $A(\mathbf{a}, \lambda) = +1$ on Tenerife and $A(\mathbf{b}, \lambda) = -1$ on La Palma. We denote this joint probability by $P(A(\mathbf{a}, \lambda) = +1, A(\mathbf{b}, \lambda) = -1)$. If we admit any possible outcome, we write $P(A(\mathbf{a}, \lambda), A(\mathbf{b}, \lambda))$, knowing, of course, that for different actual outcomes we may have different probabilities. If the occurrences on the two islands are statistically independent of each other, which they are not but we pretend for the moment that they are, then probability theory tells us that we can use the product rule

$$P(A(\mathbf{a}, \lambda), A(\mathbf{b}, \lambda)) = P(A(\mathbf{a}, \lambda)) \cdot P(A(\mathbf{b}, \lambda)). \qquad (9.1)$$

If this product rule applies, then the Bell-type inequalities follow immediately, as we will see in a moment. But what about the known dependencies and correlations? What about the correlated pairs that are sent to the two islands carrying the information denoted by λ? Bell takes care of this correlation in an elegant way by introducing conditional probabilities P, meaning probabilities that are valid when a particular correlated pair is just being detected. One denotes such probabilities by symbols such as $P(A(\mathbf{a}, \lambda)|\lambda_i)$ or just $P(A|\mathbf{a}, \lambda_i)$, meaning we have a probability that A takes on a certain value for that specifically given λ_i and setting $\mathbf{a}$. If nothing else than that λ_i gives rise to correlations for that particular measurement, then the product rule can be applied for the conditional probabilities and we have

$$P(A|\mathbf{a}, \lambda_i, A|\mathbf{b}, \lambda_i) = P(A|\mathbf{a}, \lambda_i) \cdot P(A|\mathbf{b}, \lambda_i). \quad (9.2)$$

Assume now, and this is as we know a big assumption, that each λ_i occurs about equally often with each setting. This assumption can actually be derived from Bell's other assumption that the λs and the settings are independent mathematical variables. Then, as we know, we can sort the measurement results into products using the same λ_i as Bell did in his previous proofs. Now we can form the Bell-type sum of the conditional joint probabilities that contains a Vorob'ev cyclicity:

$$P(A|\mathbf{a}, \lambda_i)P(A|\mathbf{b}, \lambda_i) + P(A|\mathbf{a}, \lambda_i)P(A|\mathbf{c}, \lambda_i)$$
$$- P(A|\mathbf{b}, \lambda_i)P(A|\mathbf{c}, \lambda_i). \quad (9.3)$$

Because all probabilities must be larger than or equal to 0 and smaller than or equal to 1, we obtain from elementary algebra

$$P(A|\mathbf{a}, \lambda_i)P(A|\mathbf{b}, \lambda_i) + P(A|\mathbf{a}, \lambda_i)P(A|\mathbf{c}, \lambda_i)$$
$$- P(A|\mathbf{b}, \lambda_i)P(A|\mathbf{c}, \lambda_i) \le +1, \quad (9.4)$$

which is again a Bell-type inequality, only now written for conditional probabilities instead of outcomes. This result has a simple interpretation. We can regard the functions $A(\mathbf{a}, \lambda_i)$, etc., all as different "unfair" coins that fall with different probabilities on heads $(+1)$ or tails (-1). Then no matter what these different probabilities are, the Bell inequality is valid.

This possible comparison to different coins is the reason why so many situations can be constructed that always fulfill a Bell-type inequality. A typical story is that about Dr. Bertlmann (see pp. 126–143 of (Bell, 2001)), a character of Bell, who likes to wear socks of different color. Bell assumes in that story that if one sock of Bertlmann is pink, the color on the other foot is not pink. Bell proceeds then to show by a complicated procedure about washing socks at different temperatures that the correlation values obtained for quantum particles cannot be explained by any process with socks. On looking closer, one finds that this is not a new proof but just one that relies on Eq. (9.4), which is valid for coins and colored marbles and, of course, also for colored and flavored socks.

In a publication from the year 1980, Bell told a story of time-dependent correlation in the two cities Lille and Lyon (see pp. 88–90 of (Bell, 2001)) and introduced an explicit effect of space-time. By explicit, I mean that he did not only suggest that the experiments be performed using clocks at large distances from each other, in two distant cities. Bell also considered the direct role of time in additional ways, in ways we know from many experiences. For example, he suggests that the weather in the two cities might be somewhat correlated and, therefore, people in both cities may not watch TV as much on sunny Sundays. These are precisely the type of correlations that Walter and I thought were crucial: independent of what is sent out from a source to the two cities, there may be other correlations that are not related to what the source sends.

Thus, Bell used times and dates in this proof and included them as an element of physical reality still denoted by his λ. However, he did not extricate himself from the erroneous assumption that λ and the settings are independent mathematical variables. His "proof" contained also one more problem: an incorrect use of the concept of conditional probabilities. Bell follows his original intuition that λ can be "anything." It can be good weather on Sundays and anything of that time-dependent kind. However, now with the introduction of conditional probabilities and Bell's use of the product rule, we have a problem as mentioned already previously. Replacing λ_i by the word "anything" we obtain

$$P(A|\mathbf{a}, \text{anything}, A|\mathbf{b}, \text{anything}) =$$
$$P(A|\mathbf{a}, \text{anything}) \cdot P(A|\mathbf{b}, \text{anything}). \quad (9.5)$$

Insert for "anything" a clock time-period or just a clock time and you have no correlations during that period or at that time. Insert the word "everything" and the statement claims complete independence. One simply cannot push concepts such as conditional probability too far by using ill-defined generalizations. If one does, one ends up in logical circles. As stated, however, Bell's use of both λ and the settings as independent mathematical variables is already lethal, and the further twists of generalization are actually not even necessary to refute his proof.

9.2 Local Beables

In his collected works (see pp. 50–60 of (Bell, 2001)) Bell introduced also the concept of "local beables." "Beable" combines the two words "be" and "able" and has been used without the word "local" in the physics literature before and after Bell's work. Bell defined local beables by taking recourse to the spirit of Bohr. Bohr stated, "It is decisive to recognize that, however far the phenomena transcend the scope of classical physical explanation, the account of all evidence must be expressed in classical terms." This sentence of Bohr means in its essence that no matter how far removed the quantum phenomena are from what we can recognize with our instruments and senses, all the evidence must be expressed in terms of the results that we obtain with our instruments, all the evidence must be presentable as a written record of events. As an aside, it is the author's belief that any meaningful probability theory must define events as they are physically defined in Einstein's special relativity and mathematically by the events in Kolmogorov's framework.

Reformulating the above paragraph in terms of Bell's mathematical notation, local beables are the outcomes of Bell's functions A for a given element of reality λ_i. These outcomes can, no doubt, be recorded by instruments following an EPRB measurement. The only object that is not well defined that way is Bell's λ_i that leads to the actual outcome. As we know by now, Bell defined λ_i as an element of reality that corresponds in essence to elementary particles such as electrons, photons, or quarks but does not include their dynamic interactions with the particles of the measurement equipment and corresponding space-time effects. (For the experts I would like to point here to the work of Gerard 't Hooft, who carefully distinguished variables, beables, and changeables ('t Hooft, 2006).)

If we wish to talk about the objects of nature and their dynamics, the totality of all elementary particles and all collections of them and their relation to each other, we need to introduce the additional elements of reality that enable the scientist and all of us to relate the objects and to predict some of their future relations or correlations. Any such prediction of future recordable events involves by necessity relations in space-time. Can these additional elements of reality be included in the set of local beables that Bell has defined? The answer is no, definitely not! The reason for this resounding no is again that Bell's beable algebra assigns the same λ_i to all settings.

9.3 Bell's Proof and "Conjectured Sampling"

Everything seems to circle about the use of the same λ_i or at least same frequency of occurrence of a given λ_i for all setting pairs. During our correspondence with Mermin an interesting fact surfaced that was related to this problem. Mermin emphasized in his publication (Mermin, 2004) that one cannot prove Bell's inequality the way Bell did for "actual outcomes of actual experiments." He further stated that affixing labels, such as our space-time labels, makes it impossible to talk sensibly about Bell's theory, because Bell's work is only about "a certain conjecture" that has implications for actual outcomes.

I did not understand how a conjecture could have necessarily valid implications for actual outcomes. Mermin also did not define this conjecture with mathematical precision. He just implied that, because one can calculate all the outcomes for all settings and a given λ_i, one can write the inequality as a *requirement* for actual experiments, in spite of the fact that only one setting pair can be chosen for the measurement of any given actual entangled pair of quantum particles.

Walter could, at first, also not make any sense of Mermin's statements. How could imagined measurements with calculated outcomes have implications for actually performed measurements and experiments? We then "translated" Mermin's words, in an attempt to gain a better understanding, in the following way: one does, of course, not perform measurements on one given entangled pair with all the different setting-pairs that are listed in Bell's inequality. However, that same λ_i that describes the given entangled pair for a given pair of settings will also be encountered and sampled for all the other setting pairs of the inequality over many measurements. Thus, these calculated outcomes involving the same λ_i can be assumed to actually occur and contribute to the average.

According to this logic, unperformed experimental results can be included, because in the long run they will be encountered anyway, and therefore a conclusion about the average of actual experiments can be drawn. If this is what Mermin meant, it is similar to Tony's argument that only one result needs to be measured, because the others are "possessed" by the quantum particle. Walter explained to me that conjectures as those of Mermin and Tony cannot be proven without further assumptions. As explained above, this guarantee that the λ_i are sampled about equally with all setting pairs cannot be given for EPRB experiments.

The guarantee can be given for functions on one probability space, i.e., random variables (see Eq. (7.1)). However, A, B, C, and D are, in general, not random variables on one probability space, because the corresponding measurements cannot be performed all at once for the same entangled pair. The identity of the measurement machinery for equal settings at different time or space-time coordinates cannot be guaranteed either. From our work (Hess and Philipp, 2004), later shown in more detail and with mathematical precision (Hess and Philipp, 2006), we knew that if the Bell inequality (or CHSH inequality and similar types of inequalities) are valid, then indeed Bell's functions can be defined on a probability space and Mermin's and Tony's conjectures are correct. If Bell's inequality is violated, however, then A, B, C, and D can certainly not be defined on one probability space. Then no set of λ_i (or of the mathematically abstract ω of Eq. (7.1)) exists of which the A, B, C, and D can be functions and, therefore, Bell's functions do not exist in the first place and no "sampling rule" can be applied to them.

It was thus our belief that Mermin and also Tony went in a logical circle. Under ordinary circumstances, when the Bell inequality is valid, Bell's functions are random variables on one probability space and one can, therefore, predict that one will sample (about equally) all the λ_i with all the setting pairs if one just measures often and long enough. Mermin and Tony, in contrast, used that rule of sampling as if it would be always valid to start with and proved Bell's inequality on this basis assuming the universal validity of their sampling rule.

However, in case of a violation of Bell's inequality, no such functions as A, B, C, and D exist and no such "conjecture" about sampling can be made when working with these particular functions. Of course, possible outcomes can then also not be calculated with Bell's functions because these functions do not exist. Mermin, Tony, and all proponents of Bell's theory did abandon the existence of Einstein's elements of reality altogether but did not abandon their sampling rule. They also did not realize that an infinite set of different functions that include space-time effects by using a separate index st_i may exist. They did not realize that these functions may depend on elements of reality λ_i that also may now depend on space-time. These new functions (and there are infinitely many of them) can always be used to remove any Vorob'ev type of cyclicity and, therefore, do not follow a Bell-type inequality, which is thus rendered irrelevant.

Chapter 10

Last Work with Walter

Und nennt die Guten, die um schoene Stunden
vom Glueck getaeuscht vor mir hinweggeschwunden.
Sie hoeren nicht die folgenden Gesaenge . . .

—Goethe, *Faust* I

Free translation:

Not favored by Fortuna's grace,
they rest now at a different place
and do not hear the words that follow . . .

Before our publication of (Hess and Philipp, 2004), my inaugural
paper for the National Academy of Sciences, Walter and I organized
a meeting at the Beckman Institute of the University of Illinois
to present all of our ideas and to discuss them with the experts.
The meeting was a disappointment for me, because not many
of the scientists who had criticized us came. Tony was present
and discussed his opposing views. I do not remember any other
discussions that we had not heard before, and the presentations that
Walter and I gave were mathematically too complex. Our adversaries
did not like what we said; only our friends did. Andrei Khrennikov
and also Louis Marchildon were present and very supportive.

Einstein Was Right!
Karl Hess
Copyright © 2015 Pan Stanford Publishing Pte. Ltd.
ISBN 978-981-4463-69-0 (Hardcover), 978-981-4463-70-6 (eBook)
www.panstanford.com

10.1 A Close Call of Nature

I have no other recollections of what happened at the meeting, and I did already suffer from heart problems that distracted me and made me nervous. Not long after this meeting, I was hospitalized and received some stent implants, because of an almost complete coronary occlusion. These stent implant procedures were essentially done in one day and I recovered quickly. Within a week, I was up and about, and began working again. As mentioned, the manuscript that we had submitted originally to *PNAS* had been rejected by a board member. We rewrote it to make it more understandable and *PNAS* had to accept it because, I believe, no inaugural paper was ever rejected. I am sure the board member who rejected our first version would also have rejected this second one if that was possible at all. With the publication of this paper much of our work on Bell's inequality was in print and we tried to decide how to proceed with our research. Walter said he wanted to publish all of our detailed refutations of Bell and mentioned to me that he could probably persuade one of his colleagues in mathematics to sponsor our manuscript for publication in a mathematical journal, which indeed happened that way (Hess and Philipp, 2006).

My work was impeded by further heart problems and after a second hospital stay and stent implantation in 2005, I started to think of early retirement and stopped working on Bell-related problems for a while. It was difficult enough for me to fulfill my many other duties at the university, including the teaching of a class. I was teaching a quantum mechanics course for engineers and this course gave me some opportunity for further thoughts on the foundations of quantum mechanics.

Sylvia was also getting ready for my retirement and prepared our vacation home in Hawaii, which was to become the permanent residence. Walter was not happy about our decision to move from Illinois to Hawaii. "This will hurt our collaboration," he said. It is much more difficult to work together with the Pacific ocean in between and with you being at the beach all day. I assured Walter that I was still very much interested in our work. Sylvia had supported me all her life and had moved to wherever I found a position that I liked. Now her time was finally coming to pick the place, and I loved Hawaii too and hoped that the mild climate would be great for my health.

Walter worked with even greater intensity on the refutation of all known "proofs" for the Bell inequality. He also helped prepare my retirement symposium, which was to be held in the Beckman Institute. During all these preparations we received an invitation to a conference in Leiden, a city in the Dutch province of South Holland. Leiden has a very well known university that was made world-famous by Hendrik Lorentz, one of Einstein's idols and colleagues who had worked out, among many other things, the way time and space coordinates transform for moving systems, the so-called Lorentz transformation. Heike Kamerlingh-Onnes, the discoverer of superconductivity, was also professor in Leiden. The conference was organized by a number of well-known physicists and mathematicians interested in the foundations of physics and probability, including Theo M. Nieuwenhuizen, Vaclav Spicka, and Andrei Khrennikov and was scheduled for the summer of 2006, shortly after my retirement symposium in Illinois. Both Walter and I felt well prepared to accept the invitation and we were looking forward to present our Bell refutations. We also were looking forward to renew our friendship with many of the attending scientists, including Andrei Khrennikov and Marlan Scully. The famous Nobel Laureate Gerard 't Hooft was also scheduled to give a presentation and we were informed that he would present new arguments against Bell's position.

Thus, while Sylvia was preparing to bid farewell to our friends in Illinois and doing the arrangements for our move to Hawaii, Walter and I were traveling to Holland. Walter had received an additional invitation from his mountaineer friends in Austria and Germany to come and commemorate some of his major climbing achievements and to visit some climbing tours that were named after him. He was planning to fly to Austria a day before the conference in Leiden ended.

10.2 Conference in Leiden

The conference in Leiden was very well organized and we met many interesting scientists. Unfortunately the weather was bad and the walk from the hotel to the conference center was frequently disturbed by lighter or heavier rains. This turned out to be an impediment to friendly discussions during the walks that would otherwise have taken place. There was a conference participant who did talk to Walter and me and showed great interest in our work. Unfortunately we had not done our homework on who he was. He did ask questions that gave away his intimate knowledge about the Bell inequality and particularly about EPRB experiments. His name was P. M. Pearle, and it should have rung a bell. He was attending the conference in company of his wife and, one day when I came back from the conference center, totally drenched from the rain and quite unhappy about the weather, he introduced me to his wife and invited me to have dinner with them. I declined, because I feared I would come down with a cold and just wanted to go to bed early. Walter was also not available that day, because he was together with some of our friends. I think Pearle took my unwillingness to have dinner as a lack of interest in talking with him, and we never got together at the conference. I know now that I could have greatly benefited from his knowledge about a whole area of "Pearle-type loopholes" for violating Bell's inequality. He had already written, more than 30 years earlier, the milestone paper titled "Hidden-Variable Example Based on Data Rejection" (Pearle, 1970). In this work he proposed a very broad way of violating Bell's inequality by just not taking all the data but instead rejecting some for more or less natural reasons. This area is still of great importance, and Bell's followers are still trying to close such so-called loopholes.

I was very interested in hearing what Gerard 't Hooft had to say about the Bell inequality. I caught him in front of a blackboard in the conference center and presented a short review of our work to him and invited him to our presentation. He was interested in our work and made a few very positive statements. Later he also wrote me a nice e-mail about our manuscript for the proceedings of the conference. Gerard is one of the greatest living physicists, and he had received the Nobel Prize for his work on "electroweak interaction." He also promised to present his views about Bell's work in his lecture, which both Walter and I attended.

Gerard mentioned in his presentation the technique of cellular automata as a possible tool of his new and deterministic description of quantum phenomena. Cellular automata were also favored as a tool for understanding nature and quantum mechanics by the well-known mathematical physicist Steven Wolfram, the father of the mathematics software MATHEMATICA. Interested readers can get familiar with cellular automata by using the examples and exercises of MATHEMATICA. Physically speaking, cellular automata models work on the basis of nearest-neighbor interactions. All the simulation space is discretized into a large number of cells and each cell interacts with the nearest neighbors on the basis of a simple rule, such as giving a particle or energy to the right neighbor. Thus, instantaneous influences at a distance are ruled out to start with. Any model of nature that is based on cellular automata has no need for such influences. Wolfram has convincingly demonstrated that this seemingly simple framework can be used to understand most processes in nature.

In his presentation, Gerard showed the possibility of deterministic theories, such as hidden variable theories involving Einstein's elements of reality, underlying quantum theory. He mentioned "some sort of cellular automaton" or a classical system of continuous fields as a possible basis to start with. Instead of quantizing such systems in the usual way, he considered what he called a "pre-quantization." He did not modify the physical system with schemes depending on Planck's constant h, but rephrased them in a language of deterministic theories that is suitable for quantum theoretical manipulation at a later stage. Thus the primary theory is deterministic, and all the great advantages of quantum mechanics can be built on top of it. I loved the presentation and was convinced that Einstein would have loved it too.

My own presentation went over well and was appreciated by several colleagues. I particularly remember interesting discussions with Theo Nieuwenhuizen and Andrei Khrennikov. Both Theo and Andrei recently published important contributions resonating with our views. The fiercest opponents of our work, however, did not attend the conference, and those who were attending our presentation did either not wish to discuss things openly or did not take us seriously; no comment was made.

At the conference dinner we tried to sit close to Gerard. However, the acoustics of the dining room was such that one could not have any sensible discussion. Walter and I enjoyed the great dinner and then returned with Andrei to the hotel. Walter excused himself early, because he had booked the flight to Austria the next morning. This was the last time I saw him.

10.3 Back in Illinois, Packing for Hawaii

Walter and I still had some written correspondence while he was in Austria and I back in Illinois. Sylvia was frantically packing up for our move to Hawaii, and I did not help her much, partly because I was busy cleaning out my office at the university and finishing business there, partly because I was pretty inept to deal with such a big problem as moving house. In my defense I can only say that I was too sloppy for Sylvia's orderly mind, and I threw away things too easily. Our vacation home in Hawaii was fully furnished. Nevertheless, Sylvia liked to take much of the nice Illinois furniture to Hawaii. Some of the pieces she sent to our daughter, who had married Kris Calef, a great guy, and lived with him in California, and others to our son, who had married Lisa a very nice lady from Kiwi-land and had moved to New Zealand. In Hawaii we were thus in a central location, at least as far as our closest family was concerned, except for my mother in Austria.

In between cleaning out my office, I sent a first draft of the paper to Walter for the Leiden conference proceedings. We had finished about half of it before traveling to Leiden. Walter in turn sent me the draft of the paper, written in German, that we had also started before Leiden. As mentioned, one of Walter's friends had promised that our work would be published in mathematical reports of the University of Goettingen. We could not get this work published in the United States in spite of trying with several journals. Somehow word had spread that *PNAS* had rejected our manuscript and only printed some version as my inaugural paper that they could not reject. One of the journal referees wrote: "They published an inaugural paper in *PNAS*, that's nothing! They should try *Physical Review*, or any other journal of significance, and they will be rejected." Indeed, we were.

On July 19, the phone rang at about 2 p.m. Sylvia was still packing and I was on the computer doing my e-mail. It was Ariane, Walter's wife, telling us that Walter had died of a heart attack a few hours earlier. We learned later that Walter was hiking with his mountaineer friends in his beloved Austrian mountains, and we saw photos of the hiking tour. It was a steep climb and most of Walter's friends were about 20 years younger. On the way back from the hike, close to their parked car, Walter sat down on a fallen tree and did not move. One friend asked him whether he was OK, and Walter answered, "Ja." This was his last word. A heart attack had taken my friend. Both Sylvia and I were very sad. Ariane flew to Austria to claim the body. Walter was cremated and his ashes distributed to his loved ones, including four children, a son and a daughter from his first wife Barbara and two sons from Ariane, his second wife.

This devastating news added to our stress related to the planned departure to Hawaii. My best friend and physician, Robert Basler, helped us with sleeping aids and consolation. He too had loved Walter and he was sad in addition, because we were leaving. Things became extremely hectic. We planned to sell our beautiful home in Illinois and were putting it on the market. The house was designed by Sylvia and specially built for our needs. It was beautifully located on a lakeside across a golf course. Neither Sylvia nor I really wanted to sell it, but I knew that it would be too much work for Sylvia to maintain two homes. So we had put it on the market with the secret hope that no one would buy it. The house was not on the market for more than four days. We had to move out by September 2006, so we started looking for a moving company. I was for hiring an expensive company so that we could lean back from now on and let them work. But this was against Sylvia's frugal grain and her wish to control all of the packing herself. She hired a low-budget moving company from the Internet that left a lot of work for her to do, not to speak of the many little problems that upset her.

I just buried myself in my work and finished the paper for the Leiden proceedings. I tried to make it as good as I could, which was not the best, with Walter's mathematical genius missing. I sent a copy of the manuscript to Tony, letting him also know that Walter had died. Tony wrote a short note that at least, having done what he liked, Walter died happy. A much longer subsequent note had a lot of criticism about our manuscript. I had tried to treat the measured clock times as random variables, which is possible because the actual measurement times are taken at random. I did this to show that even if one does this, Bell's proof runs into contradictions. Of course, I did not think that time could be seen, in general, as a random variable. Quite to the contrary, I always maintained that time needed special treatment. Tony had altogether problems with regarding some of the physical quantities involved as random variables, particularly the settings. It was clear to me that Tony thought that our approach may be correct in a mathematical sense, but irrelevant. After assuring each other that we disagreed, we assured each other of mutual respect and friendship. Tony wrote: "We are not going to quarrel now, however acute our academic disagreements may be—at least not if I can help it!" This was our last e-mail exchange.

A few days later, I received a package that Walter still had sent from Austria. It contained the finished German manuscript that he had submitted and that was later accepted for publication (Hess and Philipp, 2006). I could not even look at the manuscript at the time; it created too many emotions. I knew Walter's manuscripts were usually perfect and let the publication go forward. I read it only now while writing this book. Toward the end of the manuscript, Walter had added a statement that right from the start Bell and his followers used unproven assumptions that were necessary and sufficient conditions to prove the Bell inequality. Then they added physical conditions such as Einstein locality that were not necessary anymore for the proof, but that were blamed later to be the reason for the failure of EPRB experiments to agree with Bell's inequality. I believe now that Walter was correct not only for the special proofs that he considered then, but also for any other of the existing Bell-type proofs. I will return to this point in Section 13.1, and I am convinced that Walter's suspicion can be raised to a general conjecture that will prove true in all future investigations.

At the planned date in September, Sylvia and four workers from the moving company finished packing our household effects. The last pieces were put in the big container at 3 a.m. the next morning, by which time the work finished. In the rush, the movers broke a few pieces of Sylvia's nicest porcelain, causing some additional heartbreak during a sad farewell. Then we left our home of more than 30 years. Robert and Renate Basler, our very best friends, brought us to the airport. I looked down with tears in my eyes as the small airport became even smaller and off we went to a new life.

Chapter 11

Intermission

Life is a journey on a wooden raft in troubled waters,
not on a ship with sunny quarters.

We had purchased the house in Hawaii as a vacation home in 1999 and had always enjoyed being there. By the time we moved in permanently, we knew Kailua-Kona, the town of our residence, quite well. We also had made a few new friends. Nevertheless, such a move entails many changes and adjustments. Sylvia took the brunt of the burden after our arrival in Hawaii and ran the household and everything related to it while letting me work on my paper for the proceedings of the Leiden conference. She readied the house for our Illinois furniture, which meant that much of the vacation furniture had to go. Thinking of the situation now, it was very selfish of me to put so much of the burden on Sylvia.

Having finished and submitted the paper to Leiden, I hoped that Sylvia would stop working so hard, because the Illinois furniture was not to arrive for a few weeks. Unfortunately that was not to be, and I still blame myself for letting happen what happened then. A seemingly nice man rang the doorbell one day and introduced himself as the owner of a company that built Hawaiian lava walls.

Einstein Was Right!
Karl Hess
Copyright © 2015 Pan Stanford Publishing Pte. Ltd.
ISBN 978-981-4463-69-0 (Hardcover), 978-981-4463-70-6 (eBook)
www.panstanford.com

He had noticed that our property was steeply declining at the ocean-view side and pointed out that by building a wall and moving in rocks and dirt we could extend the property by 2000 square feet of useful area. He mentioned that this could be done for less than $30,000. I did not want to hear anything about it, because I thought that Sylvia had too much work as it was. He assured us that this would not cause any more work for us and his company would take care of everything. The machines that were needed could be easily driven in from the mountain side of the house. He also pointed out that the prices would probably go up pretty soon, because many people were building new homes. I still did not want to listen, because I thought that Sylvia and I would have a lot of work when the furniture from Illinois arrived. Sylvia actually wanted to have that extension of our property and said that, in her opinion, the furniture would take another month and arrive late, as everything does in Hawaii. She worked on me and I finally gave in and we started the wall.

When the machines arrived, it became clear that they were too big to be driven in as planned. There were a few palm trees in the way and they needed to be removed. After all of that was done and the machines could get in, the front of our house looked like a battlefield. I had paid the first installment of $10,000 for the wall and one could soon see the outline of the foundation as well as the first few lava stones on top of it. Then the wall builder came to talk to me and asked for $20,000, which were needed to complete the wall. He told me that he had 10-plus children to feed and he could not continue to build without the advance. It was stupid, but I did it. I gave him the money, after which I did not see him anymore; he left the island. Fortunately his brother promised to stand by their company's contract and finish the wall; they did not do so, however, without taking additional payments.

While work on the wall was going on, our furniture from Illinois arrived at the least opportune moment, exactly when the whole house was in dust and wall-workers everywhere. The furniture movers delivered all the bigger pieces in the different rooms where they were supposed to end up and put the remainder, boxes full of pots, glasses, books, and lots of "stuff," on the lanai (the Hawaiian porch). Sylvia started to work on everything while I mostly watched in despair. I did help her with the books after she had placed, with the help of Thomas, our gardener, our big bookshelf structure from Illinois along the wall of the living room. By the time I had all the books sorted and distributed on the shelves, Sylvia had taken care of the rest of the furniture, the kitchen, the glass shelves, and everything else that had to be done to create a livable space.

We both were happy when things were in place, but Sylvia was exhausted. I could see it on her face. She did recover, though, after a few days and we both were planning a relaxed morning with breakfast in bed. We had just had coffee when the house and all its contents started shaking. We realized immediately that this was an earthquake and jumped out of the bed and ran outside to the front of the bedroom, where we stood horrified, holding each other, and saw the house shaking wildly. There was the noise of glass breaking inside and also in all the neighboring houses. The shaking took place for about one minute, an eternity for us, with an absolute quiet to follow. The lights had gone out, because of the power failure, but we could see the disaster as the sun had just risen. The kitchen floor was five inches high, littered with glass and broken porcelain. Our Waterford collection was reduced by half. The books had mostly fallen off the shelf and had smashed a few pieces of Sylvia's decorations in doing so. We had not yet recovered, when a strong aftershock occurred. As we learned later, the earthquake was 6.7 on the Richter scale and the aftershock 5.7. The epicenter was only 20 miles away. Such a quake supposedly happens only once every 200 years in Hawaii and is not related to the active volcano, the Kilauea, but to the fact that the extinguished volcanos, 13,000 feet high mountains, sink back into the sea slowly but surely.

I really admired Sylvia and the way she took this disaster. She just said, "We had too much stuff anyway," and started with the cleaning. I believe I was not too much help for this either, although I tried my best. We had not quite cleaned up yet when we had to leave for Illinois to attend Walter's memorial. Walter's wife, Ariane, had asked me to organize a memorial at the Beckman Institute and to be master of ceremonies. The flight from Kona, Hawaii, to Urbana in central Illinois takes about 12 to 15 hours. Robert and Renate picked us up at the airport and brought us to the hotel. They would have wanted us to stay in their house, but the hotel was right across the Beckman Institute, where the memorial took place. It was a nice affair with many of Walter's friends attending. Cynthia Haymon, a friend of Ariane and well-known opera singer, performed Rachmaninov's "Vocalize" at the end, and everyone was in tears. I had taken sedatives to get through the ceremony and was very sad for the whole week, as was Sylvia. Our friends, particularly Robert, took care of us and Robert helped with further prescriptions of sedatives and sleeping pills. Soon the week was over and, with renewed pain, we had another farewell and took off to Hawaii again. In the plane, Sylvia lost consciousness for a few minutes. I called the flight attendant and she in turn found a doctor on the plane, a cardiologist. He pressed on Sylvia's chest and she woke up soon afterwards. The doctor did not find any other reason for concern at the moment and I thought it was just a brief loss of consciousness because of all the exertion that Sylvia had had and because of her usually low blood pressure. I did not realize that this was a portent of a future disaster and, using the powers of denial, had a drink and we arrived in reasonably good spirits in Kona.

The disaster started on November 24, 2006. Sylvia woke up and went for her coffee. When she returned to the bedroom she was bent over in pain, laid down on the bed in a fetal position, and told me that she suffered horrible pains in the back of her upper body. We first tried some painkillers, but when the situation did not improve, I called the office of a doctor for internal medicine. He could not see us that day and Sylvia spent the night in pain. The next morning we went to the doctor's office, Sylvia driving as she almost always did. The doctor asked her about the symptoms and after a few minutes sent us to the hospital to get a CT scan. He told us that Sylvia did not want to have what he suspected, a problem with her aorta. The emergency room staff of the Kona hospital assigned a bed to Sylvia and we waited for about three hours. A doctor stopped at Sylvia's bed once and told us that he would do the scan but did not think it was absolutely necessary. When he finally did, he returned quite upset and informed us that Sylvia had an aortic dissection and had to be airlifted to Honolulu, because nothing could be done in the Kona hospital. The plane for the airlift was ordered and Sylvia told me to go home to get whatever was necessary and then to fly with her on the ambulance plane to Honolulu. I rushed home and grabbed my portable computer and whatever I thought I would need in Honolulu and for the hospital and returned to Sylvia. She was sitting in bed very composed and without pain due to morphine injections and talked on the phone to our daughter Ursula, who promised to come as soon as possible. Sylvia had also called our son in New Zealand, and he was very concerned and upset that he could not be with us at that moment. As always, Sylvia was more concerned about me than about herself and wanted our daughter to come quickly, mainly to take care of me. Then the ambulance plane arrived and brought us to Honolulu, where they did another CT scan and confirmed the diagnosis of a downward aortic dissection, a very dangerous and long rip of the inner aorta lining that created aneurysms and a second channel, a so-called false lumen, for the blood.

The Honolulu surgeon asked whether or not he should perform surgery, but we followed the advice of our doctor friends from Illinois and decided to wait for stabilization and afterwards to find someone who could stabilize the damage further with a stenting procedure. They could not do such a procedure in Honolulu. Sylvia had to stay in the hospital because of the immediate danger and also because of the pain. She was on a morphine drip. When she was released 10 days later she was in bad shape and needed still larger doses of strong analgesics. We returned to our home in Kona and Sylvia went on hospice care, as one of the Honolulu doctors had suggested. Her recovery was very uncertain and she could get the pain medication more easily through hospice. We went into what we called the "survival mode" and tried to survive just each day at a time. Our daughter and son-in-law had joined us immediately in Honolulu and Ursula helped for the first few weeks as much as she could. She then returned briefly to California to pick up my mother in Los Angeles on her way to Hawaii from Austria. They landed in Hawaii on Christmas Day and Ursula stayed for two more weeks, now taking loving care of both Sylvia and my aging mother. She did everything possible for Sylvia in the most thoughtful and loving way and by mid January things started to improve gradually.

The recovery was very slow, but in the spring of 2007 Sylvia finally was well enough to travel to the Mayo hospital in Scottsdale Arizona which had doctors with experience in aorta stenting procedures. We picked Mayo in Arizona, because both Sylvia and I had been Mayo patients in Rochester, Minnesota, and were very impressed by the Mayo doctors. We opted for their Arizona hospital, because the travel was much easier from Hawaii, only a six-hour direct flight. We were fortunate to get the attention of Dr. S. Money, their very capable surgeon for Sylvia's type of problem. Dr. Money not only is one of the top surgeons of the country but also has a very kind and warm personality. He performed the complicated procedures that took almost a week and involved exploration by angiography, a kidney-bypass surgery, and finally the stenting, all performed with the greatest skill and most modern tools in the fall of 2007.

Robert and Renate had joined us to give us their support in those difficult times and we all stayed for a while in the house of Renate's friends Vicky and Ken Dippold in Scottsdale, including our daughter Ursula and her husband Kris. They all bestowed their warm friendship and love on us and helped me get through this difficult time. Sylvia was fighting for her life, mostly sedated and phasing in and out. Dr. Money explained all that was happening and Robert was continuously on my side and helped with his warm friendship and medical knowledge. He and Renate did not leave Sylvia's bedside, except maybe for a short lunch or dinner, with one of them always remaining there.

Sitting in the hospital during that week, I had ample of time to reevaluate my life priorities. My work on EPRB and Bell was now at the lower end of the scale of importance while I was waiting for the news from the surgeon. It did not matter to me whether Einstein was right or wrong, or whether or not I had made some contributions to that question. In fact, I did not think much about science for another six months. It took a long time, but then Sylvia finally recovered and things normalized. In the spring of 2008 I started to have the peace of mind to do some work again.

Chapter 12

A New Beginning

Starting all over again, it's gonna be rough.
As sung by "IZ" in Hawaii

In my first attempts to get back to work, I tried to find criteria for the type of experiments that could always be described by functions on one probability space and, therefore, could be modeled for sure by Kolmogorov's probability theory. My answer was: one can find a common probability space for all types of experiments that can be modeled by *two* functions, for example A_a and A_b. It does not matter whether or not these experiments involve quantum effects. This fact provided a big hint that quantization (involvement of integer numbers) by itself was not the culprit for Bell-type conundrums. The problems started when the experiments were so constituted that one was tempted to model them with *three or four functions*, such as A_a, A_b, A_c, A_d, that permitted to form a *Vorob'ev cyclicity*. As we know, it is then possible to form a Bell-type inequality that contains the cyclicity and is based on it. In case of a statistical violation of the inequality, one cannot find a probability space for these three or four functions, and one needs to introduce many more functions, such as $A_a^{st_i}$ with $i = 1, 2, 3, \ldots$ to describe the

Einstein Was Right!
Karl Hess
Copyright © 2015 Pan Stanford Publishing Pte. Ltd.
ISBN 978-981-4463-69-0 (Hardcover), 978-981-4463-70-6 (eBook)
www.panstanford.com

experiments without contradictions or influences at a distance. If one has a different space-time coordinate for each different function, the Vorob'ev cyclicity is automatically removed because all the terms of the inequality are now in principle different.

In addition, I worked out a number of necessary and sufficient mathematical conditions for the Bell inequalities to be valid, conditions that did not require any physical assumptions such as Einstein locality and that Bell had implicitly made without clear justification. I believe I made many valid points in my new manuscript and included also a few findings that Walter and I had published in German (Hess and Philipp, 2006). I tuned the language of the paper down in order to avoid annoying Bell's followers, and pointed only to inconsistencies of Bell's proof that I could unmistakably identify.

12.1 Incremental Steps and Another Shot from the Pulpit

I was pretty proud of my accomplishment without Walter's help after the long break in my work and sent the manuscript to the *Proceedings of the National Academy of Sciences of the USA* (PNAS). As an academy member I had now the privilege to choose the referees myself, and I chose Andrei Khrennikov and Louis Marchildon, the experts I knew from the conferences in Sweden.

Both referee reports were very favorable, but the Academy informed me that my work was rejected. Such rejections were, as I was informed, relatively rare and the overwhelming percentage of member submissions was and is usually accepted. This fact had been criticized by some as an unjustified privilege. However, the tradition of most European academies is to print practically anything members submit. In the heydays of German physics, manuscripts and corrections were often hand-carried between authors and printers to speed up the process. Refereeing was not even necessary and still is not in some of the academies. I had a very positive referee report of two well-known experts. As mentioned, Andrei had edited several books on the subject and written at least two himself, and Louis had written a well-known quantum mechanics book with a chapter on the Bell inequality. Both scientists have an impeccable reputation and would not have let me get away with any flaw that they detected. This, however, was not enough for the editor in chief of *PNAS*. He may have thought that I was a typical case of a successful scientist who believed that he still could solve difficult problems and then overreached. Be all of this as it may, a *PNAS* board member whose name was not revealed to me wrote the following:

> In recent years the field of "quantum foundations," which was previously a sort of Cinderella subject, has become a respectable sub-field of physics and has nucleated an identifiable intellectual community. It is important to appreciate that the mere familiarity with the formalism of quantum mechanics (QM) and fluency in its standard application does not automatically qualify one as a member of this community.... Like most vibrant intellectual communities, the field of quantum foundations has its schisms. A major one is between a majority who feel that the work of the late John Bell on what is sometimes known as "quantum nonlocality" is one of the most profound results of physics, and a minority, I would estimate comprising perhaps 10% of the whole, which feels that this work is either wrong or somehow trivial. I belong to the majority camp; I suspect that Dr. Hess would identify himself as belonging to the minority (though no doubt he might object to my characterization of it).

Indeed I do object to this characterization. One cannot divide physics into strict partitions of quantum foundations, high-energy physics, many-body quantum mechanics, and the like and claim a majority, because physicists of one subfield think the same way and are then defined as the only experts. When it comes to influences and actions at a distance and so-called quantum nonlocalities, there is a majority of physicists all across many areas of expertise, including some of the best known and still living stars (see Section 12.2.2), who are against instantaneous influences at a distance. In addition, I do not think that one should suppress the publications of a minority, even if it truly were a minority. The work of the greatest, of Einstein, Newton, and others, would never have been published if such a rule were invoked. Of course, I realize that I am not of that class. I also know, however, that my manuscript contained no obviously disqualifying mistakes and should have been favorably considered for publication.

The *PNAS* board member wrote further: "Over the last few years Dr. Hess, sometimes with collaborators, has published a series of papers which attempt in some sense to refute or minimize the work of Bell. My response to most of these papers, which I suspect is typical of the majority, is that these papers are not so much wrong as flogging a very dead horse, and thus of little interest."

It amazes me that my objections, and identification of actual mathematical mistakes in Bell's work, were so summarily dismissed and labeled a dead horse. I would have expected that a *technical* summary needs to be given to justify rejection of the work of a colleague in the Academy by a board member. However, the board member played another unusual card and attacked the referees:

> Dr. Khrennikov is certainly in the minority camp, while I don't know what position, if any, Dr. Marchildon takes. In any case, while both are perfectly reputable physicists, I think it would be a stretch to describe either of them as a "famous expert"; at least in the foundations of quantum mechanics.... Dr. Khrennikov, while he has certainly been active in this field for quite some time, I suspect he would not make most people's list of (say) the top two dozen people in the field.

Of course, Andrei Khrennikov would not make the list if only those scientists vote who form the "majority" and are against his work. The board member obviously did also not appreciate the interdisciplinary character of the research in question, and Andrei Khrennikov's very significant contributions on the side of mathematics and probability theory. Louis Marchildon is one of the best quantum physicists I know. He is very precise, critical, and open-minded.

There were also two new referee reports commissioned by the board member. One of these referees admitted that (s)he was somewhat out of her/his depth: "I confess that I have not understood all of this manuscript, but I acknowledge that it may be partly my fault, my lack of familiarity with Kolmogorov probability spaces." In her/his final analysis, however, the referee proceeded to criticize my manuscript, arguing as follows: "The Bell theorem is a direct continuation of the EPR reasoning. From their assumptions (local realism + the correlations predicted by quantum mechanics are correct), EPR *prove* the existence of additional elements of reality. Then Bell calls lambda these elements, uses locality again, and derives his inequality, which turns out to be contradictory with some predictions of quantum mechanics. Hidden variables are never assumed, just local realism, this is all." I quoted these sentences of the referee verbatim, because they show several major misconceptions about Bell's work.

EPR did not *prove* the existence of elements of reality; they only claimed that either there exist spooky influences or that quantum theory is incomplete and elements of reality exist. Bell introduces his elements λ and *assumes* that these elements form one variable of a function that also depends on the equipment settings as second *independent* variable. These unjustified assumptions form already necessary and sufficient conditions for his inequality to be valid. Bell does not need to use "locality again" to prove anything further. The term "local realism" and consequences that follow from its definition are discussed in detail in Section 13.4, which explains further misconceptions related to this referee's comments.

There were also some justified comments by this and a second new referee that, in my opinion, could have been taken care of by small corrections. The *PNAS* editor in chief, however, rejected my manuscript. I was hurt and wrote a blistering letter that I would not deal with *PNAS* in the future, a promise that I broke soon afterwards, because I wanted to help some younger colleagues to get their important work published. Be all as it may, I am convinced that publications by a member of the National Academy of Sciences in *PNAS* should not be suppressed on the basis of the objections of a subgroup of scientists, who claim to represent a "majority."

12.2 Quantum Nonlocality or Just Uncertainty?

Some of the words of the *PNAS* board member rendered me sleepless for some time. Particularly his sentence containing the phrase *"quantum nonlocality"* is one of the most profound results *of physics* really bothered me. I knew, of course, that Bell himself believed in quantum nonlocality. He presented this view in a publication shortly before his death in 1990.

This paper of Bell was directed against the words of the famous physicist H. B. G. Casimir, one of Wolfgang Pauli's students. Casimir had given the following example for correlated events without influences at a distance. Boil an egg and watch at the same time a running clock. The egg is coagulating as the clock is ticking, "yet the coincidence of these two unrelated causal happenings is meaningful, because I, the great chef, imposed a structure on the kitchen." Bell titled the paper containing his counterargument "La nouvelle cuisine" (see pp. 216–234 of (Bell, 2001)) and included a section, "Quantum mechanics cannot be embedded in a locally causal theory." I knew from the Leiden conference that Gerard 't Hooft did not agree with this statement and I remembered the statements of Murray Gell-Mann against quantum nonlocalities and did a limited literature survey that demonstrated to me the following. While many physicists believed that Einstein was not entirely correct and that Bell's work was very important, most of them were opposed to instantaneous influences at a distance. Below I am describing a few highlights of my survey. The expert reader who wishes to know more is referred to the many publications in the proceedings of the conferences organized by Andrei Kherennikov (for example (Khrennikov, 2002)) that often tell a more detailed story.

12.2.1 *The Uncertainty Principle and EPRB*

It is difficult to assess with complete certainty how the original adversaries to Einstein's views—Bohr, Heisenberg, Born, Pauli, and others—would have reacted to the EPRB experiments of Aspect and Zeilinger as well as many others and whether they would have then proposed quantum nonlocalities, as Bell did. From reading their works, I am convinced that they would not have accepted any instantaneous influences at a distance. What they were mainly opposed to were Einstein's opinions about the Uncertainty Principle.

Whether or not the EPR paper has something definite to say about the Uncertainty Principle itself is still a subject of discussion in the modern literature. Possible "exceptions" to the Uncertainty Principle based on EPR are discussed by (Ozawa, 2003), but go beyond any elementary presentation at the level of this book. It is true, however, that whenever it comes to measurements with different settings on both sides of the EPRB experiments, the language and reasoning of EPR becomes less clear. They write, for example (on p. 779 of their paper), "We see therefore that, as a consequence of two different measurements performed on the first system, the second system may be left in states with two different wave functions. On the other hand, since at the time of measurement the two systems no longer interact, no real change can take place in the second system in consequence of anything that may be done to the first system." Note that now two different measurements are invoked with the first system, i.e., with one particle, which complicates things when one talks about the Uncertainty Principle.

For the EPRB experiments as performed by Zeilinger's group, one can "translate" the EPR statement from above as: "We see therefore that, as a consequence of two different measurements performed on the photon in Tenerife, the second in La Palma may be left in states with two different wave functions." We can detect now an important problem with the EPR formulation (as was pointed out to the author by Robert Griffiths): one cannot perform two different measurements on the same photon in Tenerife. Einstein himself was not happy with this formulation, as he almost immediately admitted to Schrödinger.

If, however, we do not deal with the Uncertainty Principle by itself, but just the basic tenet of the Copenhagen interpretation of quantum mechanics and the question about spooky influences, one can "sanitize" the sentence of the EPR paper in the following way. Assume that the distance that the photon propagates to arrive at the measurement point in Tenerife is slightly shorter than that of the photon measured in La Palma. Then, if we take at time t_1 a measurement in Tenerife with the entangled pair being prepared in the Bell state of Eq. (7.4), there is according to the Copenhagen school a 50% chance that we obtain an H-type measurement result and in that case know with great certainty that in La Palma they will obtain a result corresponding to the H state. Had we measured a short time later (at t_2) with a vertical setting in Tenerife, we would have had again a 50% chance to obtain a V-type result. That being the case, we would have known, again with great certainty, that the result in La Palma will correspond to the V state. Now one can reason, as EPR did, that no real change can take place for the La Palma photon because of any of these measurements in Tenerife and, therefore, some element of reality must be involved that determines both the outcomes in Tenerife and La Palma.

Our space-time-dependent elements of reality will, as we know, do the trick. The so translated EPR reasoning, however, is now more twisted, because it involves assumptions about unperformed experiments. This problem together with Bell's work provided the reason for the opponents of EPR to abandon the idea of elements of reality. If, however, Bell's work is not sufficiently general as explained in this book, then EPR appears in a totally different light and the correlations of measurement results with the same settings on both sides clearly point to the existence of elements of reality. How else would one explain that the measurement outcomes on both sides are entirely random but completely correlated? The only other explanation would be influences at a distance, just as Einstein said. Arthur Fine has also been giving extensive discussions relating to these points (Fine, 1986).

Bohr himself did not believe in spooky influences and had only one defense against EPR's subtle argument. He stated, "There . . . is no question of a mechanical disturbance of the system under investigation during the last critical stage of the measurement procedure. But even at this stage there is essentially a question of *an influence on the very conditions which define the possible types of predictions regarding the future behavior of the system.*" As far as I understand this sentence, it points only toward the predictive quality of quantum theory and does not explicitly deny the existence of Einstein's elements of reality. The difficult point that EPR invoked unperformed experiments, and that Bohr may have objected to also, has been discussed above and is going to be addressed again in Section 13.4.

Schrödinger took the EPR side and stated that Einstein had caught the Copenhagen school by their coat tails. It is very telling that one of the fathers of modern quantum theory makes such a statement. The Copenhagen school saw it naturally just as a ploy of their competitor Schrödinger. The famous Wolfgang Pauli did not like Schrödinger's view and noted that the introduction of the elements of reality would just destroy the effectiveness of the quantum theory.

The mathematician John von Neumann later took a stand similar to that of Pauli. He added, however, the first mathematical proof that such elements of reality could not exist. Walter had looked at this proof, and although he was a great admirer of von Neumann, he thought that this proof was abstruse and outdated, because it did not refer to modern developments of probability theory. Be that as it may, von Neumann's proof was destroyed by a number of scientists, including Bell himself. Both von Neumann and his friend Eugene Wigner have defended the idealist position that Einstein's ball is not in either box. I think it was this "idealism" that led to their opposition to Einstein.

12.2.2 *Recent Reactions*

Reactions in recent times about instantaneous influences and Bell's position have been given from the viewpoint of quantum theory in its most modern versions, from the viewpoint of modern classical probability theory, and from mixed perspectives.

12.2.2.1 Gerard 't Hooft

Gerard 't Hooft ('t Hooft, 2006) developed a mathematical basis for a deterministic theory that may underlay quantum mechanics. As described previously, he suggested that one may suspect, among other possibilities, an underlying system of cellular automata. Such a system is typically based on interactions of nearest neighbors only and excludes, therefore, influences at a distance. Cellular automata have received the attention of mathematicians and have been shown to be suitable for formulating many occurrences and correlations found in nature. 't Hooft presented an even more general case. Instead of quantizing such a classical or "ontological" system in the usual manner, 't Hooft introduces a pre-quantization, which means a rephrasing in a language suitable for quantum manipulations at a later stage, but not yet based on a new constant (such as

Planck's constant h). His work soon mentions the role of the concept of time and the Hamilton operator acting on vectors of Hilbert space. His deep understanding of quantum theory cannot be captured here. However, the fact that he finds a deterministic theory underlying quantum mechanics possible cannot be stressed enough. He certainly comes down on Einstein's side in the EPR discussions.

12.2.2.2 Gell-Mann, Griffiths

Gell Mann called influences at a distance "flapdoodle" (Gell-Mann, 1994). He thus clearly expressed his contempt. Griffiths, Gell-Mann, Hartle, Omnes (GGHO), and others developed a new framework of interpretation and application of quantum mechanics that is now known as the "consistent quantum theory." It shows how instantaneous influences at a distance can be avoided in quantum theory (Griffiths, 2002).

GGHO introduce, as already described in Section 7.3.3, consistent histories. The word "histories" gives, of course, a hint that the concept of time is used as an entity that labels the order of events. However, they do not use space-time alone to describe the order of events. They add a quantum version of the concept of context. "Context" refers to other conditions that are affiliated to the existence of some "properties" that describe a given situation. As an example consider a property such as "a gold bar in your safe." It then makes sense to talk about "in your safe" only if you have a safe. One therefore says that the property "gold bar in a safe" depends on the property "safe." One can formulate such dependencies for very general "properties" such as a certain spin polarization of a photon and other quantum states. GGHO have shown that in a formal way and have used the context in their logical descriptions of quantum theory in addition to the time ordering. Einstein's space-time elements of reality are thus replaced by a time-ordered context. Whether or not this approach goes beyond Einstein's elements of reality and space-time is, of course, a tantalizing question. Fortunately we do not need to answer this difficult question when searching for possible Einstein local reasons for violations of Bell's inequality.

I have identified as a necessary and sufficient condition for the Bell inequality to be valid that the functions A that Bell has introduced are functions of two independent variables: the settings **a**, **b**, **c**... and λ. Without this assumption, Bell's proof fails. The fact that λ and the settings cannot be treated as independent variables has been discussed in great detail throughout this book. Griffiths also points to Bell's functions and addresses this fact, because he introduces consistent histories and thus does not permit different settings for the same time coordinate.

The reason given by Griffiths is, one cannot use different settings at the same point of history, based on the fact that quantum mechanics associates with such different settings mathematical operators that do not "commute". Explained in lay terms this means the following. Mathematical operators are mathematical expressions or objects that act on the quantum states that are represented as vectors in a so-called Hilbert vector-space. The expression that these operators do not commute means that the order in which these operators are applied makes a difference and the outcome is not necessarily the same if we apply the operator corresponding to setting **a** first and that corresponding to setting **b** second or vice versa. This fact relates to what we explained above about properties and context. There may be different contexts depending on what type of measurements are performed in what sequence.

This reasoning leads in principle to the same result that we derived from a consistent application of the space-time picture, namely that the same λ cannot be used for all setting pairs of the Bell inequality without making additional assumptions, and we have, therefore, no problem accepting it. There are, however, the following two facts to consider in favor of the "pure" space-time reasoning that we have used all along. (i) The involvement of commuting and noncommuting operators brings with it the involvement of quantum theory and its connections to experiments, which is an additional complication to an already complicated problem. (ii) Einstein himself believed that noncommutation does not play any role in his EPR reasoning. He stated in his letter to Schrödinger (dated June 19, 1935) that the incompatibility, meaning the noncommutation of the "quantum observables," did not matter at all to him, because his general reasoning did not depend on this fact at all. If we believe in this statement, then it must be possible to derail the proof of Bell by the proper use of space-time and without reference to the operators of quantum mechanics. The author is convinced that this is indeed what is accomplished in this book.

The following facts lend further support to the latter statement. Noncommutation by itself is not a sufficient reason, and may also not be a necessary reason for violations of Bell's inequality. The inequality is valid for an infinite number of different settings and corresponding noncommuting operators and thus noncommutation is not sufficient for violations. It also may not be necessary. Consider the performance of many different EPR experiments for different settings with results that all can be recorded at once. We could, for example, have thousands of measurement stations on both Tenerife and La Palma. Then all measurements are performed in two different cities on differently generated entangled pairs and all the operators commute according to the rules of quantum theory. A violation of Bell's inequality may still occur in such experiments, thus making noncommutation a condition that is not necessary to obtain violations. However, because such experiments have not been performed, my remarks about noncommutation are currently only a conjecture.

12.2.3 *Loopholes, Prisms, Filters, All against Nonlocalities*

There have been many papers on the "loopholes" in Bell's work, meaning circumstances that get around Bell's assumptions and, therefore, permit violations invoking instantaneous influences. It is particularly one assumption that is difficult to achieve experimentally, which is that either every particle pair emanated from the source is indeed registered by the measurement equipment, or at least that a "representative" sample of pairs is encountered in the measurements that leads to the same long time average. Pearle was the first to present a very general description of this particular loophole in his milestone paper (Pearle, 1970). Inefficiencies of the photon detectors could supply such a way out. Fine (Fine, 1986) also described theoretical possibilities to achieve violations of Bell's inequality that he called "prism models." These prism models may be related to experimental detection efficiencies too. Fine stated explicitly: "The question of efficiency, however, is also relevant to the survival of the prism models."

Hans De Raedt, Kristel Michielsen, and coworkers have developed very interesting computer simulation methods that used what they call "filters." What happens in their simulation is that, under very reasonable assumptions, some of the correlated pairs with special properties are filtered out (not included in the averaging procedure). Their filters also encompass models such as that of Pascazio (Pascazio, 1986) that considers time delays and has been discussed in Section 5.2.3. A comprehensive review of the work involving filters is presented in (De Raedt and Michielsen, 2012), which also covers a whole range of interesting quantum experiments other than EPRB. Fine calls filter-type models "synchronization models". I do not like this particular name, because synchronization smacks too much of synchronizing certain settings with certain source-events, which leads to violations of Bell's inequality with greatest ease. Such synchronizations are not admissible, because of the random change of settings after the entangled pair is emitted from the source. Fine did, of course, not refer to such synchronizations but just referred to the fact that the coincidence in the time window of the measurement on the two islands is of great importance. As we have outlined in Section 2.2.1.1, this coincidence is determined by computer clocks and "filters" out particles that are not within a certain time window. The EPRB measurements, thus, depend always on some coincidence-in-time filtering that makes it very difficult to exclude filter effects altogether. All of these loopholes present significant alternative explanations of violations of Bell-type inequalities.

We have concentrated all along on the question whether Bell's proof is valid even if all emitted entangled pairs are measured, and the author's answer is that it is not, because Bell made additional assumptions that cannot be defended and that have been explained by the role of space-time. The argument that I have already presented can be further sharpened, as is explained through the knowledge of the following sections that was obtained in collaboration with Hans De Raedt and Kristel Michielsen. The author feels fortunate to have received, several years ago, an e-mail from Hans that discussed their work related to EPRB. From this first contact a friendship and collaboration developed that led not only to a number of results adding to what I have already presented in the previous sections, but also to a new and clearer explanation of objections to Bell's proof, all based on the work of Boole. In fact, I do not think I could have written this book without that collaboration. Hans sent me a copy of Boole's paper from 1862 that contained Bell's inequality. Only then, Bell was not yet born.

Chapter 13

The Inequality and Boole

Nihil novi sub sole.

—Vulgate Bible

There is nothing new under the sun.
Hypotheses non fingo.

—Newton

I contrive no hypotheses.

My first contact with Hans and Kristel was, as just mentioned, through e-mail exchanges. We commiserated about how easy it was to get something published that indicated a violation of Bell's inequality and how complicated it was to publish anything opposed to the gospel of Bell. I also brought Tony Leggett's work with Anupam Garg to the attention of Kristel and Hans and we started a detailed correspondence and very soon a collaboration that resulted in a visit of Hans and Kristel in Hawaii.

The work with Hans and Kristel resolved and clarified several important facts and resulted in several publications. Hans, Kristel, and I became friends, and with them and through them the work that I had started with Walter took off toward new horizons.

Einstein Was Right!
Karl Hess
Copyright © 2015 Pan Stanford Publishing Pte. Ltd.
ISBN 978-981-4463-69-0 (Hardcover), 978-981-4463-70-6 (eBook)
www.panstanford.com

13.1 Boole's Variables, One-to-One Correspondence

Boole's work on statistics and probability (Boole, 1862) is an important precursor to Kolmogorov's probability theory, which was discussed in Section 7.3.2. Boole was very aware of the logical difficulties that may arise when one considers events of the physical world and one wishes to speak about the likelihood or chance of such events in a mathematical-logical way. His pathbreaking solution was to dissect the events one deals with into more elementary events that leave only two alternatives open. When we consider events of nature in a probabilistic way, he stated, we need to dissect these events into the "ultimate possible alternatives which they involve." Such alternatives could be "true" or "false," while for our case, the EPRB experiments, these ultimate alternatives are either a detector click or no detector click following the photon–polarizer interactions. Boole introduced variables that can assume these alternative values and comprise the mathematical abstractions of the experimental facts. These abstractions have been called the members of the sample space in Section 7.3.2.

Boole stated that we need to be able to deal with these sample space elements in a logical fashion, meaning we need to be able to apply the logical connections AND, OR, as well as NOT to them. If we wish in addition to deal with numbers, because we are so familiar with them, then we have to introduce functions that map these events onto numbers. For the EPRB experiments, we have denoted these functions by $A_{\mathbf{a}} = \pm 1$, etc., and these functions need therefore to follow the rules of integer numbers. How do we know that all of what we have been doing is correct, that all our abstractions do correspond to the real experiments, that the events can be connected logically by AND, OR, NOT, that the functions of these events follow the rules of real numbers and, therefore, lead us to the long-term averages of these experiments correctly and precisely? This is the question Boole had posed to himself in a section of his paper titled "Determination of the conditions of possible experience."

Boole showed that one must have a so-called one-to-one correspondence of the mathematical abstractions to the actual experimental outcomes. This one-to-one correspondence is of great importance, and without this correspondence the whole logical network that we are building when describing experiments by probability theory becomes nonsensical. One can see that from the following example of our number system that we use for counting. Let's assume that we are counting oranges. Then we "create" maps or functions that assign to any set of oranges a number. One orange is mapped to the number 1, two oranges to the number 2, and so forth. For each amount of oranges we have exactly one number and for each number we have the corresponding set of oranges. This is called a one-to-one correspondence of elements of reality and mathematical abstractions, which are in this case the numbers that we count with. If we only make one mistake in that correspondence and, for example, assign to 100 oranges also the number 5, then the whole system loses its usefulness. Merchants would only buy sets denoted by the number 5, hoping that once in a while they would get 100 oranges. All this is, of course, trivial. However, if we deal with objects of nature that we cannot identify as easily as oranges, objects such as photons and electrons, then the one-to-one correspondence becomes a less trivial requirement.

For each possible EPRB experiment with the two possible outcomes, we need mathematical variables that are associated with the experiment. If we have denoted the mathematical variables by $A_{\mathbf{a}}$, $A_{\mathbf{b}}$, $A_{\mathbf{c}}$, then we need to make sure that each experiment can indeed be represented by these variables, and if they are functions of λ, as Bell assumed, then we need to associate with each experiment a certain λ. For example, we may have for the tenth experiment a λ_{10} and setting $\mathbf{b}$ in Tenerife and obtain the outcome $A_{\mathbf{b}}(\lambda_{10}) = +1$. How could it happen that, in spite of Bell's careful choice of events and functions, the events do not follow an algebra of logic and the functions A do not follow the algebra of real numbers? Well, we can have, for example, more complicated factors that influence the experiment and that Bell did not know, and this lack of knowledge prohibited the assumed one-to-one correspondence of the abstractions with the experiments. This is best explained by an example. Here is one that Hans, Kristel, and I concocted. Naturally it follows exactly the pattern of EPRB experiments, but now we have patients instead of photons and doctors instead of polarizers.

13.2 Possible Experience for Doctors

Consider two doctors, one in Tenerife and one in La Palma, who perform examinations of patients with a certain kind of disease at random but equal dates of the calendar year. The doctor in Tenerife examines patients from Spain and enters into his notes $A_a = +1$ if a patient from Spain is diagnosed positive for the illness and $A_a = -1$ if diagnosed negative. He also examines patients from England and enters for them the note $A_b = +1$ if positive and $A_b = -1$ if not. The doctor in La Palma also has patients from England and again enters for them in his notes $A_b = +1$ or $A_b = -1$. She also has patients from the USA and enters for them $A_c = +1$ if they are positive and $A_c = -1$ if not.

The doctors are convinced that neither the date of examination nor the location (Tenerife or La Palma) has any influence on the illness and therefore denote the patients by their countries of origin, because the origin matters in their theory of the disease. After a lengthy period of examination they combine their results by forming products (of results from Tenerife and La Palma respectively) and find the following averages:

$$\langle A_a A_b \rangle + \langle A_a A_c \rangle - \langle A_b A_c \rangle = +3 \tag{13.1}$$

These, as we know now, violate Bell's inequality. They further notice that the single outcomes of A_a, A_b, and A_c are randomly equal to ± 1. This latter fact completely baffles them. How can the single outcomes be entirely random while the products are not random at all, and how can one obtain the average of $+3$, which is the maximum possible outcome? After lengthy discussions they conclude that there must be some influence at a distance which leads to opposite outcomes for the English patients on the two islands. They also believe that this strange influence leads also to random fluctuations for the outcomes. This explanation works, as the dedicated reader can find out. Just use the same result as that of the English in La Palma for all the A in the equation, except that you must choose the opposite for the English in Tenerife. Thus, influences at a distance can explain this strange result of the examinations.

However, there are more reasonable ways of explanation that have nothing to do with influences at a distance that make the examination of the English on Tenerife always the opposite of that on La Palma. For example, we can have a time dependence and an island dependence of the illness that could arise from different water supplies for different days and on different islands. On even dates we could have $A_a = +1$ and $A_c = +1$ on both islands while $A_b = +1$ on Tenerife and $A_b = -1$ on La Palma. A nasty person could, for example, put something in the water of La Palma that hurts only the English-born, because of some features of their constitution. On odd days all signs are reversed, because of the deeds of other nasty persons. Obviously for doctor exams on random dates we have then the outcome that A_a, A_b, and A_c are randomly equal to ± 1, while at the same time the result of the equation above is $+3$. We need no deviation from conventional thinking to arrive at this result, because now we just add the coordinates of the islands and the date, that is, the space-time coordinates in our equation. For example, we can add the superscript "te" for even days and "to" for odd days in Tenerife, and "lpe" for even days and "lpo" for the odd days in La Palma. Then we obtain $A_a^{te} A_b^{lpe} + A_a^{te} A_c^{lpe} - A_b^{te} A_c^{lpe} = +3$ for the even days, and a similar equation for the odd days, with the e in the superscripts replaced by an o. A random choice of even and odd days leads then to the average shown in Eq. (13.1). The reason that this result is possible without invoking any influences at a distance is that the **b** setting (the English patients) are treated differently for Tenerife compared to La Palma and different on even and odd days. This difference removes the Vorob'ev cyclicity that we had described in Section 8.3, introduces the randomness, and also results in a new one-to-one correspondence between experiments and the mathematical abstractions. This is exactly what Boole had in mind: if there is a violation of an inequality, then there is something fishy with the mathematical abstractions that were used and one needs to look for new ones.

The inequality that Boole actually used in his example was not precisely identical to that of Bell but very similar. In our current notation, Boole's inequality would read

$$A_a A_b + A_a A_c + A_b A_c \geq -1 \qquad (13.2)$$

Note that the inequality of Boole is different from Bell's and has a $\geq$ sign (larger or equal) instead of a $\leq$ sign (smaller or equal).

It also has three plus signs and no minus on the left-hand side. It is easy to show, however, by elementary algebra that Boole's and Bell's inequalities are mathematically equivalent. The interested reader is also encouraged to insert all possible values of ± 1 for the functions A to verify that the inequality is indeed valid. Boole came up with this inequality 100 years before Vorob'ev cyclicities and Bell's inequality were discovered. This fact demonstrates clearly that such inequalities are not necessarily related to some particular physics. They are always valid if their mathematical abstractions make sense. If they do not, and if the one-to-one correspondence of abstractions and actual experiments is more complicated than assumed, then the inequalities can be violated.

There is, of course, an additional fact that Bell has emphasized and that complicates matters, which is the equality of outcomes for equal settings. In the above example this would mean that the English patients would have to show the same outcome when simultaneously examined on both islands. Proponents of Bell could say, "There you go. Then you need some influence at the distance after all. How, otherwise, could it happen that the English patients show always different results except when they are simultaneously examined on both islands?" This is, of course, one of the reasons why Bell found it impossible to explain violations of the inequality without influences at a distance. The requirement of equal outcomes for equal setting and random outcomes for a given measurement station is also the Achilles' heel of Bell's work, because equal settings do not appear in his inequality and require different measurements with different space-time coordinates compared to the measurements for non-equal setting pairs. These different space-time coordinates open then possibilities to violate Bell's inequality while still permitting equal outcomes for equal settings and, therefore, invalidate Bell's proof.

As explained previously, a time window during which the entangled pair is detected needs to be created. This can be done by the use of very precise and synchronized clocks on the two islands. As we know from Section 4.2, different settings on the different islands give rise to different time-dependent tables for the $+1, -1$ outcomes as shown, for example, in Fig. (4.2). Consider the case that we have made a measurement on Tenerife that is modeled by the application corresponding to a function A_b, while on La Palma we have a measurement modeled by A_c. Subsequently we perform a measurement for equal outcomes using A_b on both sides. Are the three apps corresponding to A_b all the same? Can an identical mathematical abstraction, an identical function, be used in all three cases? The answer is no, not in general. All along, we have given the reason that there may be space-time dependencies that make it necessary to attach a space-time label. Now we can see the necessity for this from both a mathematical and modeling point of view and also from the physics of the experiments.

Mathematically speaking and from the view of computer simulation, one may "align" different possible outcomes for the A_b function or application in Tenerife, depending on what setting is used in La Palma, because the measurements are taken within a given time window and the measurement timing may be different for different settings. Physically speaking, the measurements with different settings exhibit a different symmetry of the equipment and may, therefore, be different in their nature. Furthermore, any detailed theoretical understanding of EPRB experiments also needs to include the atoms and elementary particles that constitute the polarizers. The quantum particles of the polarizers are "shaken" by the incoming entangled particles. That shaking may depend on the settings of the polarizers in a variety of ways that are not controlled by the experimenter. Last but not least, environmental influences need to be accounted for, such as the rotation of the earth and the presence of electromagnetic fields and any number of other influences.

As a consequence, we need to affix a time index or space-time index to the functions to obtain functions $A_b^{st_i}$, $A_b^{st_{i+1}}$, etc., that are different for different space-time index. Naturally, if the setting on Tenerife stays the same but that on La Palma changes, then the index i that counts the measurements must change to $i + 1$. Therefore, Bell's one-to-one correspondence of functions, which counts all A_b the same is incorrect or at least not general enough. We may have to include a much larger number of functions. The same is true for any other setting $\mathbf{a}$, $\mathbf{c}$, ... This invalidates the proof of Bell, who uses the same functions for the same settings and, therefore, a very simplified one-to-one correspondence of experiments and mathematical abstractions. The author doubts that this problem can ever be resolved in such a way that Bell's proof can be restored, because of the fact already mentioned above. Whenever one likes to have equal settings and equal outcomes on both sides, an experimental adjustment needs to be made that differs from the settings that are used in Bell-type inequalities and one, therefore, needs different space-time coordinates. Thus the appropriate one-to-one correspondence of abstractions and experiments will always be a question. Andrei Khrennikov and Theo Nieuwenhuizen have recently discussed this problem from the viewpoint of context. Theo uses in his Foundations of Physics paper the word "loophole" when discussing the importance of specifying the context. He introduces the term "contextuality loophole" and stresses that it cannot be closed, because it is not an experimental complication, but based on the logical error that the context needs to be specified, as discussed extensively in this book.

Bell did not look for such a one-to-one correspondence or deficiency of it. He just assumed that all functions with equal setting are the same. As shown previously, different λs do not help, because λ cannot stand for a space-time label. Bell connected violations by experiments to the quantum nature of the experiments and nonlocal effects. We are now in a position to understand that both of these beliefs are incorrect. The introduction of different, and particularly space-time-labeled, one-to-one correspondences between measurements and mathematical abstractions provides an infinite number of possibilities to create violations. This fact considerably "softens" the barrier for Einstein local explanations of EPRB experiments that Bell derived. Violations of the Bell inequality by experiments become, therefore, of lesser basic significance while experimental deviations from the quantum mechanical result appear now in a more important light. One cannot say anymore that EPRB experiments deviate a lot from what is explainable in Einstein local ways and still a lot, but less, from the quantum theory. This was pointed out to me by Hans and Kristel, and I will return to this point in Section 13.6.

13.3 Independent Variables, Quantum Operators, and Einstein Locality

After realizing that Boole derived his inequality decades before quantum theory was developed, we can ask ourselves also the question whether quantum theory is linked to Bell's work and his inequality in other ways than discussed up to now, in ways that do not just relate to the quantum expectation values for long-term experimental outcomes. It is fairly easy to see that there is no reason to connect the basic elements of quantum theory, the quantum operators, to Bell's inequality. Einstein stated already that his reasoning and his example with the ball in the two boxes had nothing to do with commutation or noncommutation of quantum operators. Quantization itself, the fact that one distinguishes only two outcomes for the spins ($+1$ and -1), is also not necessary for one to obtain Bell-type inequalities, as we have seen from Boole's example above and from Bell's Eq. (9.4) in Section 9.1, when he discussed Bertlmann's socks. Nevertheless, Bell believed that *violations* of his inequality were of special importance for quantum theory and its foundations. We know now, however, that violations can be avoided by the inclusion of space-time and the removal of the Vorob'ev cyclicity.

There is still the question of Einstein locality and whether the experiments on Tenerife and La Palma (or any other pair of islands or places) involve some instantaneous influences at a distance. Walter's last message to me was elaborating on his suspicion that Bell's use of Einstein locality, his demand that λ must not depend on the actual setting of any of the two measurement stations, was a redundant addition and that other assumptions of Bell were already both necessary and sufficient to validate his proof.

Indeed Einstein locality is neither a necessary nor a sufficient condition for the validity of Bell's inequality. One can imagine a world of nonlocal spook that still leaves the Bell inequality untouched and valid. Einstein locality is also not sufficient to validate Bell's inequality, as we know from the example with the two doctors. One thus can see, and Walter has shown this with mathematical rigor for several specific cases in our publication (Hess and Philipp,

2006), that Einstein locality really has nothing much to do with Bell's inequality. What is it then that Bell assumed and that leads to his inequality?

As emphasized over and over, Bell used functions of two independent variables, the settings and λ. He never wrote the settings, however, as one usually writes a variable. The mathematical way of doing this would be to introduce a new symbol, say $\mathbf{j}$, that is the setting variable and can stand for all settings $\mathbf{a}$, $\mathbf{b}$, $\mathbf{c}$, $\mathbf{d}$ ... and then write the functions as $A(\mathbf{j}, \lambda)$. Bell rarely used such a notation. Instead he used the functions always with a given fixed setting: $A(\mathbf{a}, \lambda)$, $A(\mathbf{b}, \lambda)$, $A(\mathbf{c}, \lambda)$. Thus it escaped me early on that Bell was actually using a second independent variable $\mathbf{j}$ in addition to his λ.

The latter fact became crystal clear to me later when I worked with Kristel and Hans, and I realized then also that Bell was even violating conventions that were sacred in quantum theory. The independent variables used in physics since Newton were the space coordinates and time, i.e., x, y, z, t. Boltzmann and other founders of thermodynamics, the physics of heat phenomena, added momentum coordinates p_x, p_y, p_z, energy E, and mass m as another set of possible independent variables. The momentum coordinates are simply the mass m of a particle such as an atom multiplied by its velocity. This gives for the x direction $p_x = mv_x$ and similar expressions for the y and z directions.

The event of Einstein's relativity theory brought significant changes to what could be seen as independent variables. Einstein clearly showed that there was a relation between mass and energy $m = \frac{E}{c^2}$ and that the mass depends actually on the velocity. Even the coordinates x, y, z, t could not be expressed independent of the velocity of a given physical system. As everyone knows now, and as easily can be measured by the atomic clocks of our GPS system, the times that different clocks show depend on the relative motion of the clocks. One still can describe a physical system that moves with constant velocity (a so-called inertial system) by the four independent variables x, y, z, t that describe our four-dimensional space-time. Physicists like such sets of four independent variables and have also become used to other four-sets such as p_x, p_y, p_z, E. However, one cannot combine such sets. To measure both space- and momentum-coordinate, one needs two different equipment setups that cannot be switched from one to the other during an infinitesimally small time period, because of the limiting velocity of light c. Simultaneous measurement of space coordinate and momentum of a given particle is, therefore, not even possible according to Einstein's relativity. It is also not possible according to quantum theory, because of the Uncertainty Principle.

Quantum theory always emphasized the Uncertainty Principle and never treated variables such as the setting variable $\mathbf{j} = \mathbf{a}, \mathbf{b}, \mathbf{c} \ldots$ as independent of space-time variables. In fact the settings are used as labels for the quantum operators that are mathematically separated from the space-time variables that appear in the wave function $\psi(x, y, z, t)$. This is where Bell's approach differs significantly from both quantum theory and Einstein's relativity.

Bell assumes in all his many derivations that settings $\mathbf{j}$ and λ are independent variables of his functions. He deduces from this fact that any λ, for example $\lambda = \lambda_1$, occurs about equally often with any setting, for example $\mathbf{j} = \mathbf{a}$. This fact is then a sufficient reason for his inequality to be valid, as his proof shows. It is also a necessary reason. For, if one type of λ occurs more frequently with setting $\mathbf{b}$ than with setting $\mathbf{a}$, violations of the inequality can easily be constructed. Therefore, Bell assumed to start with a necessary and sufficient condition for his inequality to be valid. That condition has nothing to do with Einstein's locality condition and is physically questionable, because a generally introduced space-time coordinate can only occur with precisely one setting and not with any other.

13.4 Macroscopic Realism

Kristel, Hans, and I were now ready to scrutinize the work that Tony published with his collaborator Anupam Garg (Leggett and Garg, 1985). Their work takes up the question of quantum theorists, whether the moon is there when no one looks. Tony and Anupam did not refer to the moon but to the magnetic flux in superconducting quantum interference devices (SQUIDS). The magnetic flux can, in essence, assume only two values that are denoted by ± 1, as we did for the polarization of photons in the EPRB experiments.

Their reasoning is not dependent on any particular EPR-type of experiment and applies also to the EPRB experiments with photons. The reader not familiar with SQUIDS may wish to continue to think in terms of the familiar EPRB experiments, except that we are now dealing with magnetic flux instead of polarized photons. Tony and Anupam highlight the connection of Bell's (or actually Boole's) inequality and quantum superposition, and link this superposition now more directly to macroscopic effects such as the magnetic flux, Schrödinger's cat (dead or alive), and Einstein's exploded and unexploded gunpowder.

I follow in my description Takagi's book and his chapter the "Cat and the Moonlight." (Takagi, 1997) describes with great clarity why he does believe, with Tony and Anupam, that Bell's work contradicts macrorealism, the kind of realism that Einstein subscribed to and that is so firmly cemented in common views. "Macro" refers to the macroscopic world, the world that we see, as opposed to the world of the elementary particles and atoms, the quantum world. Macrorealism is defined by the following two statements.

MRA1 If a macrosystem under certain conditions is, whenever observed, found to be in one of two or more macroscopically distinct states, then one can assign to it, at almost all times, the property of "actually being" in a particular one of these states, even when it is not observed.

MRA2 For the macrosystem, it is in principle possible to observe which of the states it is in without affecting its subsequent behavior.

The notations MRA1 and MRA2 indicate that we are talking about macrorealism (the letters MR) and about assumptions (the letter A), and we have two of them. In spite of the fact that the two statements contain qualifiers such as "almost all times" and words that have no entirely precise definition such as "states" and "certain conditions," it is pretty clear what is meant. Using Einstein's gunpowder example, the meaning is the following. MRA1 says that there are two "states" of the gunpowder, the unexploded state and the exploded state, and that the gunpowder is at almost all times in one of these states, if we ignore a certain short period of time when it is just exploding and part of it is still unexploded. Those who believe in macrorealism assume the latter statement to be true, independent of whether or not anyone is observing the powder/explosion. Einstein, supposedly with a roaring laughter, asked one of his friends: "Do you really believe that the moon is not there when you are not looking?" MRA2 says then that our actual observation of the exploded or unexploded gunpowder makes no difference. Of course the experiments of physics are not always that clearly defined. If we watch the roulette in the Casino, we do not know whether or not there are hidden magnets that can guide the rolling ball into certain grooves. All that we observe are the groves with the numbers $0, 1, \ldots, 36$ and the ball falling into precisely one of them, giving us 37 "states" that the ball may end up in. If hidden magnets are involved, we would have a lot more possible "states" and the person controlling the magnet does have an influence on the experimental outcome.

Tony and Anupam define also variables, mathematical abstractions, that describe the measurements and that we call here A_1, A_2, A_3, all of them being able to assume only the two values ± 1. Hans, Kristel, and I could not discern any difference to Bell's abstractions except that the numbers 1, 2, 3 replace Bell's settings **a**, **b**, **c**. Tony and Anupam assume that the three indices can also be linked to times t_1, t_2, t_3 and also be used for many experiments performed in parallel. However, this would mean that one only can use three different time indices (or space-time indices if many parallel experiments are involved), which is a very severe restriction to the use of the space-time concept, both physically and mathematically. Leggett, Garg, and Takagi then state that it immediately follows from MRA1 that one can define a joint probability density $P(A_1, A_2, A_3)$ for the results of any triplet of measurements as, for example, $A_1 = +1, A_2 = -1, A_3 = +1$ and any other combination of this kind.

The existence of joint probability $P(A_1, A_2, A_3)$ represents a major step to derive Bell-type inequalities. It is known that a joint probability $P(A_1, A_2, A_3)$ exists and that A_1, A_2, A_3 are, therefore, random variables on one probability space, if the experiments can be described by precisely three "machines" or apps that operate like coins with a given likelihood for falling on heads or tails. The outcomes of the experiments with these three "machines" must be recordable in triples. This is not a trivial requirement for any experiments related to Bell-type inequalities, because they involve measurements of different entangled pairs at different times in two stations. If instead many such EPRB stations are used simultaneously, the experiments may still involve dynamic effects that depend on the different space-time coordinates, because they might involve interactions with different equipment and surroundings. Therefore, the necessity may arise to think of more than three different machines, possibly an infinity of such machines, even if they look all the same to the observer.

To complete their theory, Tony and Anupam use MRA2 to justify their use of only *three* abstractions A_1, A_2, A_3 to describe the *six* experimental outcomes listed in any Bell-type inequality. This justification is also difficult, because Bell's inequality contains each of the random variables A_1, A_2, A_3 twice and one needs to show that these pairs of functions are indeed identical. In Bell's notation this means that one can use the same λ for all terms in the inequality. MRA2 is actually insufficient to show this identity of the λs and Tony and Anupam had to add later another assumption. Not surprisingly, this assumption is related to a variation of the "restaurant problem" that we discussed in Section 5.2.1 and will be explained below.

I had very detailed discussions with Kristel and Hans about all of these facts. None of us could see a way to deduce the existence of a joint probability for the mathematical abstractions A_1, A_2, A_3 from MRA1, without making very significant additional assumptions. The concept of a joint probability is completely man-made just as the concept of the natural numbers is. The *A*s must be *functions* on a *set* of *events* that must obey some form of *algebra* as we have outlined above. Last but not the least, *a one-to-one correspondence of all the mathematical abstractions to the actual measurements must be guaranteed*. Each word or phrase in italics adds another layer of definitions and assumptions.

If one cannot rely on sense impressions alone but uses algebra, how does one find out whether indeed the events form an algebra and whether indeed the *A*s are functions on the algebra of events? Boole told us how. Form an inequality containing a cyclicity (now one may call it a Vorob'ev cyclicity) like

$$A_1 A_2 + A_1 A_3 + A_2 A_3 \geq -1 \qquad (13.3)$$

Then look and see whether this inequality is statistically fulfilled by the experiments. If not, then you have to choose a different one-to-one correspondence of abstractions and experiments.

Tony and Anupam proceeded the other way around. They maintained that the existence of the joint probability $P(A_1, A_2, A_3)$ "immediately follows" from MRA1. Then they deduce Boole's inequality of Eq. (13.3) by using MRA2. Incidentally, Takagi named Eq. (13.3) the Leggett–Garg inequality. Because their inequality contradicts quantum mechanical predictions, they deduced that quantum mechanics contradicts macroscopic reality. Thus, they went in a logical circle on the basis of the incorrect belief that they could derive the existence of a joint probability from MRA1 and to use only three abstractions because of MRA2. Instead, they should have just concluded that their mathematical abstractions did not exhibit a one-to-one correspondence to the measurements and were not elements of an algebra—all reasons to choose different mathematical abstractions.

On top of this, Hans and Kristel reexamined the work of Ballentine (Ballentine, 1987), who showed already in 1987 that the actual quantum mechanical predictions for triples, such as A_1, A_2, A_3, are different from the predictions for pairs like A_1, A_2, A_1, A_3 and A_2, A_3. Ballentine blamed this fact on a violation of assumption MRA2. The consistent history interpretation of Griffiths, Gell-Mann, Hartle, and Omnes gives that answer as well, because it clearly distinguishes between pair and triple joint probabilities (Griffiths, 2002) on the basis that the experiments are performed in a different "context."

Not without difficulties we published our work (Hess, Michielsen and De Raedt, 2009) and there were subsequent replies by Tony and Anupam published in the same journal. Tony and Anupam emphasized in their reply the additional assumption that they had introduced later, and they maintained that this additional assumption would take care of all problems. They named this assumption "induction" and gave the following example for it (using Tony's notation of Section 2.3): "The fact that a given photon 'would have' given result $+1$ had A been measured on it is not affected by the fact that it was actually B rather than A which was measured on it." This statement clearly points to the logical problems that have plagued the reasoning surrounding Bell's work and that were discussed in connection with our restaurant example in Section 5.2.1 and from the viewpoint of probability spaces in Section 9.3. It also reminded me of a discussion with Tony that took place many years ago.

Tony had explained to me that the EPR paper contained some illegitimate dragging-in of unperformed experiments. These also were mentioned to me by Griffiths, as discussed in Section 12.2.1. If Einstein used unperformed experiments in his reasoning, why should Tony not use them in his dealings with macrorealism? I believe that this problem is a very deep and difficult one and goes to the heart of Einstein's convictions. Einstein believed that science could not be based on sense impressions alone. He gave the example of the system of natural numbers. This system cannot be derived in its totality from sense impressions, although it is closely related to them. There is, however, a one-to-one correspondence between numbers and counted objects such as oranges. I believe, with Einstein, that we can sometimes go in our thinking beyond the counting, beyond sense impressions without any illegitimacy. We can talk about an infinity of numbers, for example, and we can also talk about what would have happened if another measurement was made. The problems start, however, when the one-to-one correspondence of mathematical abstractions and events is somehow jeopardized or violated. For the example of natural numbers, we need to have such a one-to-one correspondence when counting oranges.

The problems increase still further when possible outcomes that one would have measured, the results of unperformed measurements, are added to the long-term averages as if they had indeed happened. This step has been taken by Tony, Anupam, and others. Mathematically this step may still be legitimate under certain conditions. If we know that we deal with functions on a probability space, then we still may be safe. This knowledge, however, cannot be applied for Bell's functions whenever his inequality is violated. Not realizing this fact may lead to circular logic: if Bell's inequality is valid, we can use the rules for functions on a probability space. Therefore we use them and prove the Bell inequality, which in turn validates the rules. This was already discussed in Section 9.3 and was not considered by Tony and others when using unperformed experiments and introducing them into the conceptual framework of probability theory to calculate long-term averages.

The author believes that the findings presented here should end all the claims that quantum theory and macroscopic realism contradict each other. Such a contradiction cannot be derived within a precise mathematical probability framework like that of Boole and/or Kolmogorov and it also cannot be derived within a precise formulation of quantum probability as that introduced by GGHO.

13.5 What about Interference?

The readers, even those willing by now to believe that Bell's work contains serious flaws, may ask about other quantum effects that do seem to contradict Einstein. A well-known example is the two-slit experiment, the experiment that uses a foil of metal with two parallel slits in it. If one shines light on it, or directs a beam of electrons toward it, then the pattern on a screen that is observed for light (photons) or for electrons going through the slits has the following property. Cover one slit and measure the pattern of particles that one detects on the screen. Then cover the other slit and measure again the pattern and then add up these two patterns. The result is significantly different from what one obtains with both slits open. This fact continues to be true if one reduces the photon and electron number of the beam so drastically that one is certain that only single photons and electrons pass the slits at a time. How can that be? Many experts of quantum theory see only one way to explain this: the electrons and photons must somehow go through both slits when both slits are open. This is analogous to saying that if we are not looking in one of the two boxes we cannot say where the ball is and, therefore, we have a superposition of states, of the ball being in the two boxes. We now have a superposition of the photon or electron being in both slits.

One can look at this type of experiment in the following different way, too. Instead of the two slits, consider just the material that was taken out of a metal foil to form the slits. Then we have in essence two metal rods that resemble antennas. Send now a photon or electron beam toward the two metal antennas and observe the pattern of the photons or electrons on a screen after the particles have interacted with the electrons and atoms that constitute the antennas. The same phenomenon occurs: the patterns observed for each antenna separately do not add up to that observed with two antennas being present simultaneously. The careful reader can already see where we are going here, having used the word simultaneously. We somehow need to include a dynamics to explain this experiment in a complete fashion. What do the antennas, their atoms and electromagnetic fields (for experts, actually all the local gauge fields related to the experiment), do dynamically when they interact with incoming electrons and photons? Is the result of the dynamics of single antennas that are present separately at a run of experiments measuring a pattern adding up to result in the dynamics of two antennas being present simultaneously during another run? What happens to the electrons, atoms, and the electromagnetic fields of the antennas? Anyone who has learned about electricity and antennas knows that one plus one does not equal two, not for antennas. So why are we surprised that the patterns do not add up for antennas or slits?

Mathematically speaking, we have again a problem with the one-to-one correspondence of the abstractions and actual events. The events with both slits open and those with only one, the events with both antennas and those with only one, cannot necessarily be brought into a one-to-one correspondence. The existence of a Boolean algebra for these abstractions cannot be guaranteed by just attaching indices that indicate through which slit a particle might have passed. The indices must also account for how many slits were open or how many antennas present. The reason is simply the different dynamics that creates different statistical dependencies. For example, electron oscillations in the two antennas cannot be treated as independent. Thus, the author believes that we have for all of these quantum puzzles a similar problem with both the math and the physics: we simply cannot ignore the dynamics and we cannot separate the incoming particles and those of the slits or antennas and link them to *independent variables* or postulate statistical independence. Using a more rigorous mathematical approach, Andrei Khrennikov has made similar points in his book (Khrennikov, 1997) on pages 81–87. The consistent history approach of GGHO does also have a way out of the conundrum which is, however, somewhat complicated and cannot be described here. It is certainly free of any instantaneous influences at a distance.

Hans and Kristel have even presented corpuscular event-by-event simulations for Mach–Zehnder interferometers, proving that the phenomenon of interference does not have to lead to quantum nonlocalities. They and their coworker Mikhail Katsnelson expressed this fact also in a very general way. Taking a very novel point of view, they derived the quantum theory for a number of experiments, including EPRB, and were also able to derive the Schrödinger equation that forms one important basis for the theory of quantum phenomena (De Raedt, Katsnelson and Michielsen, 2013). To accomplish this, they used a general framework of probability theory called "logical inference," which contains Boole's approach that we have discussed above as a special case and is also very compatible with that of Kolmogorov. In contrast to Kolmogorov, the framework of logical inference does not use probability spaces, as the conventional quantum theory also does not.

In addition to logical inference, they use only two assumptions:

1. There is an uncertainty of the individual events whose outcome is neither considered nor predicted.
2. Certain properties of large collections of events such as the long-term average or expectation value are reproducible, meaning that small changes of the experimental conditions do not change them.

Hans, Kristel, and Mikhail need also, of course, space-time, which they introduce in a very straightforward way by postulating that classical equations that describe the dynamics of systems (such as the equations of Hamilton–Jacobi) hold *on average* with space-time coordinates describing this average. Because they can derive the quantum results on such a simple basis, they also can explain EPRB, two-slit, and other experiments without taking recourse to any influences at a distance or any assumptions of superposition of states.

Let me just mention in passing another set of experiments that seem to suggest that quantum theory is intrinsically nonlocal.

Zeilinger and his group have used entangled triples and quadruples that also have served as a "prototypes" of qubits useful for quantum computing. Manuel Aschwanden from Switzerland, who worked with Walter Philipp and me toward a master's thesis, found explicit time-dependent models that describe all the outcomes reported by Zeilinger's group without taking any recourse to instantaneous influences at a distance. The interested reader is referred to the original work of Manuel (Aschwanden et al., 2006) that is elementary enough to be studied by anyone with a good background in algebra. Manuel worked all of this out in a few weeks during his stay in Illinois and has my full admiration for it.

13.6 Experimental Uncertainties

As a consequence of my interactions with Hans and Kristel, I acquired through them a more detailed knowledge of the EPRB experiments and their accuracy and statistical significance. Kristel and Hans have studied the EPRB experiments in great detail and performed a detailed analysis of the experimental data that were published by the Zeilinger group, in particular by Gregor Weihs, and also by other groups. The careful presentation and statistical analysis of these data (De Raedt, Michielsen and Jin, 2012) was a revelation for me, and I list here only their major results.

1. The time window of the measurements is clearly important and the amount of violation of Bell's inequality that is measured depends directly on the width of the time window. Violations become noticeable at a time window width around 150 ns and increase with decreasing window size, down to the smallest window sizes of about 1 ns. Time is thus directly relevant to the experiments. For the smallest time windows, when the violations are largest, the number of measurements of particle pairs becomes very small, so that the results may not be statistically meaningful.

2. For equal settings on both sides there was a 10+% deviation from the expected equal outcome in the data presented by Weihs. This is a very significant deviation from

quantum theory that demands agreement with probability 1. As we know, without a very precise equality of outcomes for equal settings, the Bell inequality may be violated.

3. The statistical deviations of EPRB experiments from quantum theory are much larger than those of experiments that made quantum mechanics famous. The energy of photons emitted from atoms is reproducible to many digits and the quantum theory of spectra agrees with the experiments in all these digits. This accuracy of quantum theory is typical for all experiments involving electronic states in atoms or molecules. Furthermore, the so-called anomalous magnetic moment of the electron has been calculated and measured in agreement to about 14 digits. In stark contrast, the EPRB experiments exhibit deviations of 10% to even 15% between experiments and quantum theory. Such a huge deviation hints toward some basic difficulty that has otherwise not been seen in quantum experiments. In fact, Hans and Kristel showed that the known optical EPRB experiments deviate from the results of quantum theory by significantly more than can reasonably be explained by the usual uncertainties in experimental precision.

The large deviations of EPRB experiments from ideal quantum theory suggest to me that there is something missing in the basic understanding of these experiments. It also shows clearly that applications of the physics of entangled particles that require a high degree of accuracy, such as quantum computing, are currently way beyond reach.

In passing, I would like to reemphasize that the demand of equal outcomes for equal settings is a very important one for Bell's proof. The Bell inequality and that of Clauser–Horne–Shimony–Holt do not contain terms with equal setting, and experiments are often presented in the literature that show violations of the inequalities but do not discuss how accurately equal outcomes are obtained for equal settings. To start with, the term "equal settings" is not even well defined. The settings are pronounced as being equal if they are adjusted to result in the most frequent equal outcomes. Specially arranged experiments, can result in 99% equal outcomes or even slightly better, as Paul Kwiat assured me several years ago. However, as mentioned above, the data of Weihs are far from this precision. It is my personal conviction that EPRB experiments that do not continuously test equal settings during the measurement sequences with the different settings that appear in the inequality, and EPRB experiments that do not show a very high percentage of agreement for equal settings (much better than 90%) are of reduced value for discussions regarding violations of the Bell inequality.

It is clear from our theoretical objections to Bell's proof that there are many Einstein local ways to violate Bell's inequality. Bell's inequality represents, therefore, just a very "soft" threshold for the onset of phenomena that are difficult to explain in Einstein's way by invoking elements of physical reality. The significant deviations of experimental results from quantum theory appear thus in a different light. We cannot say anymore that the experiments give results that are closer to quantum theory than they are to Einstein local theories. This fact presents a problem that was highlighted by the work of Hans and Kristel and needs to be investigated in great detail in the future.

Chapter 14

Was He Right?

Scio nescio.
> —Socrates of Greece

I know that I know nothing.

There was a follow-up to the Leiden conference organized in
Prague (Frontiers of Quantum and Mesoscopic Thermodynamics),
and Vaclav Spicka sent me an e-mail with an invitation to give
one of the plenary presentations. Zeilinger was also to give one in
the same session, and Hans and Kristel told me that they would
attend and give a presentation too. Looking forward to see also
Andrei Khrennikov, Theo Nieuwenhuizen, and many other friends,
I accepted immediately and started to prepare a presentation in
collaboration with Hans and Kristel. I also wanted to go to Prague for
two personal reasons. My grandmother was Czech and I had never
been in Prague, because of the problems with the "iron curtain"
that fortunately did not exist anymore. The second personal reason
was that Robert Basler, my best friend in Illinois who had died the
previous year, had visited Prague, where he had lived in his youth,
as his last travel wish and had told me how great this city was. I also
was looking forward to visit my mother in Vienna on my way back

Einstein Was Right!
Karl Hess
Copyright © 2015 Pan Stanford Publishing Pte. Ltd.
ISBN 978-981-4463-69-0 (Hardcover), 978-981-4463-70-6 (eBook)
www.panstanford.com

to Hawaii. Sylvia had collected mileage upgrades so that I could fly first class. Going any other way from Hawaii is painful, because of the enormous distances involved; one travels around half the world. I flew from Hawaii to San Francisco and stayed there for a night, and the next day to Frankfurt and Vienna, from where I took the bus. When I finally arrived at the hotel in Prague after a total of two travel days, I was pretty tired, in spite of the fact that I had my own bed on the 380 Airbus from San Francisco to Frankfurt.

14.1 Prague

On my first day in Prague, a Sunday in July 2011, I walked from the conference hotel down to the city center. The hotel was located on one of the hills surrounding the city that was nested in the valley at the shores of a mighty river. There was a narrow road, starting as a walkway and transforming into a narrow street, winding downwards, with restaurants, bars, and churches on the side. The churches seemed all rededicated to being concert halls and each one promised performances of Mozart, Dvořák, Smetana, and other classic composers. Palaces and beautiful buildings emerge from the hazy views of the top as soon as one approaches the city limits and enters the unique center that is quite different from any other city that I know. It may not be as grand as Paris, London or Vienna, but it has a warm and engaging beauty, an inviting friendliness and a dignified impressive appearance, all at once. The streets were filled with people of all walks of life and a density that I have seen otherwise only in Tokyo. I spent the whole day walking around, having lunch and later coffee and sweets, and I barely made it back to the hotel for the welcome reception.

In the hotel I first ran into Vaclav Spicka, who welcomed me and provided conference information. Wine and aperitifs were served in a neighboring room that was still almost empty when I entered. There was only one genius-grey-haired gentleman and his wife walking in at the same time, and he introduced himself as Anton Zeilinger. The friendly face was overshadowed by darker clouds when I mentioned my name, and Anton explained to his wife that the two of us had a scientific disagreement. Both Zeilingers were about to separate themselves politely from the just recognized adversary, when I assured Anton that I had the greatest respect for his experiments and that it was only Bell's theory that I opposed. Of course, I knew that Anton had a romantic love for instantaneous influences at a distance, and I hinted toward this preference of his, joking that whether he or I would prevail on this topic did not matter: an Austrian-born person would be right anyway. I do not think that this little joke was received all too well, because Anton and his wife looked relieved when they saw some friends coming in and they turned toward them. I too turned to the incoming friends and started chatting with Hans, Kristel, Andrei, and Theo as they arrived. I also talked to some of my colleagues from Illinois who were present, including Hans Frauenfelder, Gordon Baym, and Paul Kwiat. Then I went to bed early, because my presentation was the next morning in the first plenary session.

In the morning I avoided the conference breakfast and took the coffee in the hotel's expensive restaurant as the only guest. This gave me an opportunity to recapitulate the major points I wanted to make in my talk. Then I walked over to hear the welcome greetings of Vaclav, Theo, and others. The presentation of Zeilinger followed soon. He presented the highlights of their experiments in Tenerife and La Palma and stressed the possibility or even necessity of instantaneous influences at a distance as a consequence of the statistics of these experiments. I believe he was already dreaming of experiments involving satellites and greater distances between the two measurement stations. I am not sure, however, because I thought mostly of my own presentation instead of listening to Anton.

My presentation followed and I made my major points that Bell's theory did not appropriately account for the role of space-time and was, therefore, incorrect in its conclusions about deviations from Einstein's locality. I explained the example with the doctor examinations that I had worked out with Hans and Kristel. Walking up and down the stage, I talked about one doctor doing his exam in Tenerife and the other in La Palma and explained how these exams would violate Bell-type inequalities. I also mentioned Boole's work and his almost identical inequality, and that it could be violated if there was a lack of one-to-one correspondence between mathematical abstractions and actual experiments. The applause after my presentation was nice and I expected some tough questioning of the scientists opposed to my views, particularly of Anton Zeilinger. However, none of them had questions. This vacuum permitted one of the participants to get up and give a five-minute lecture that he had anticipated much of what I said many years ago. Then the session chair pointed at the clock and did not accept any further comments or questions and started the next presentation.

I was slightly disappointed, but was happier later in the hallways, where I got a number of approving comments. One was from Gordon Baym in the elevator, where he told me that he liked my example with the doctors. Paul Kwiat on the other hand did not approve and just mentioned that still no one could show him how to play the Bell game. I said that I could explain to him why the Bell game could not be played, but he did not have time then. Turning away he mentioned that he would be free toward the end of the week and that I should try to catch him then during one of the coffee breaks. This lack of enthusiasm by a younger colleague from my home university turned me off and, after talking to Hans and Kristel about it, I decided that I would not try to explain to Paul why the Bell game could not be played. I knew, of course, that this was also one of Tony's objections.

I have described the Bell game already in Section 5.3. It is played by two players, one in Tenerife and the other in La Palma. They cannot communicate with each other and they choose their settings **a**, **b**, **c**, **d** perfectly randomly just before one of the entangled pairs arrives. The difference to the actual experiments is that the players know exactly if and when the randomly sent out pair arrives and that every sent out pair does indeed arrive and they must make a choice of $+1$ or -1 for each entangled pair. The long time averages for the measurements with the setting pairs **a**, **b**, **a**, **c**, and **b**, **c** need to violate Bell's inequality. Furthermore, for equal settings on both sides there need to be equal outcomes. Whoever can do that wins the game.

These seemingly innocuous rules of the game exclude all effects related to the time window, because the entangled pairs are, as ruled, detected with certainty, without any involvement of space-time dependent measurements. The rules also exclude any and all dynamics related to possible many body effects in the measurement equipment (polarizers), because of the demand of settings **a**, **b**, **c** that are constant during the measurement and serve as a complete characterization of the polarizers. However, any actual experiment detecting space-like separated correlated pairs cannot be performed without the use of clocks.

The challenge of the game is, therefore, as impossible to fulfill as is the very reasonable counterchallenge to perform EPRB correlation experiments without clocks or time windows being involved. After excluding space-time effects in the measurement equipment altogether, only spooky influences remain left for any explanations.

A further problem is, as mentioned previously, that the game cannot be played by probability theory, because one cannot use any general joint probability that depends on the settings of both sides. This is simply so because no player knows the setting of the other side and therefore cannot know the joint probability except if the events on both sides are completely independent. Even though Einstein local joint probabilities depending on both settings exist, one cannot use these joint probabilities, because one does not know the setting of the other side. This also means that the Bell game cannot be played by using any laws of quantum theory, even if they are Einstein local, because quantum theory is a probability theory and the quantum theoretical average for EPRB measurements depends on the settings of both sides. Tony and Paul maintained in previous discussions that this did not matter, because quantum theory may be nonlocal. Nature, however, returns a result for each measurement on each island. If Einstein is right, and no influence can propagate faster than the speed of light, there must be a way to "play" the game.

I believe that this type of reasoning led great scientists like Tony and Anton to embrace instantaneous influences at a distance, because they could not think of a way to play the game. They did never proof, however, that the game cannot be played. They also never accounted for the fact that nature does not directly "demonstrate" EPRB coincidences. These coincidences are based on measurements during man-made time windows that involve the man-made concept of space-time. Nor does quantum theory tell us how to relate the measurements of correlated pairs, and all their interactions with the polarizer particles, to the time windows. One would need to do a complete quantum theoretical calculation that includes the entangled pair as well as a time-dependent dynamical treatment of all particles constituting the two polarizers and detectors as well as their relevant interactions with the environment.

There are several ways to play the game under conditions that include the always used time window of the measurements in some fashion. As mentioned, if the measurement equipment introduces delay times that depend on the settings, then one may measure a single photon instead of two correlated ones, because one photon arrives outside the time window that is set for the particular experiments. In this case, the measurement machinery has, without any fault of the experimenter, led to a perception that one of the correlated photons has not been transmitted by the polarizer, while in fact it may have, but just has arrived too late. Examples of measurement delay time effects and violations were pointed out by Pascazio (Pascazio, 1986) and later by Gill and Larssen and others. One can also play the game with the use of general "filters" as shown by the computer simulations of Hans, Kristel, and coworkers. For each of these very specific ways of playing the game, there may be some variation of EPRB experiments that exclude that specific way that led to the violation. The literature is full of published loopholes that permit to get around Bell's theory, and of variations of experiments that exclude the given loophole. In addition, one has to deal with the large experimental "errors" that are present in all the EPRB experiments. The question remains whether there exists a method by which the game can always be played and that cannot be excluded by some more or less complicated reasoning and different experimental arrangements. This question is obviously very difficult to answer and has not yet been answered.

I believe that I have shown in this book that the possibilities for violations based on space-time effects are so numerous that one can go on arguing forever, as long as one does not have the definitive knowledge of how nature actually does it. Einstein's space-time is flexible enough to find an infinity of Einstein local ways to violate Bell-type inequalities, because one can find an infinity of mathematical abstractions that are more complex than those of Bell. Unfortunately, it is impossible to disprove instantaneous influences at a distance. Such "spooky" influences cannot be disproven by the very definition of the word spooky.

After this first day in Prague, and the reasonable success of my presentation, there was a whole week of conference that I could enjoy without worries. I had several long walks and discussions with Hans and Kristel and enjoyed lunches and dinner with them. The conference featured also a large number of unusually elaborate receptions, including concerts and dining as well as cocktails and wine. Vaclav and the conference organizers had found a way to combine science with lots of fun, actually a very ingenious way.

Science award ceremonies often suffer from a lack of attendance, simply because scientists do not travel far just to see some of their colleagues getting honored. This is particularly so if the honors go to scientists that have left their country of origin a long time ago and just return because their home country honors them. I remember situations of high government officials awarding medals in almost empty halls. The organizers of the conference and the award committees of Prague got around this problem by combining conference receptions and concerts with award ceremonies of the Czech Republic. The conference participants wined and dined enthusiastically at these ceremonies which provided them with luxurious surroundings, gourmet food, and great music. I got to know a number of churches and palaces and will not forget these beautiful evenings. At the end of the award ceremonies, I usually walked back to the hotel with Hans and Kristel, up the steep hill, I slightly inebriated, and all of us convinced that Einstein was right.

After the conference I took a bus to Vienna to visit my mother. Hans and Kristel drove me to the bus station and saw me off. I spent a few days with my mother and my cousin plus family in Baden and enjoyed the hot springs of the town. Then I received a phone call that Sylvia had an arterial blood clot and was hospitalized. I rebooked my flight immediately and returned to California, where Sylvia had visited our daughter Ursula and son-in-law Kris and was now in the hospital. Fortunately, the surgery to remove the blood clot went well, and I returned soon afterwards together with Sylvia to Hawaii. There I had finally time again to think about what conclusions could be drawn from the conference in Prague and what Einstein would have said about it. Hans, Kristel, and I also submitted a paper summarizing our thoughts (Hess, De Raedt and Michielsen, 2012).

14.2 Gott Wuerfelt Nicht

I have discussed throughout this book Einstein's opposition to instantaneous influences at a distance, to quantum superposition, and (in a qualified way) to the Uncertainty Principle. Einstein had a problem with the use of probability in the laws of nature and particularly in quantum theory. He preferred a "theory whose fundamental laws make no use of probabilistic concepts." The interested reader may also wish to consult the detailed quotations and explanations by Arthur Fine in his well-known work about EPR and Bell (Fine, 1986). In a more popular way, Einstein summarized his position by the famous one-liner "Gott wuerfelt nicht" (God does not play dice).

News reports quoting Einstein usually claim that Einstein despised probability and thought that everything in nature happened with certainty. This is, of course, not true. Einstein was a great master of probability, made major contributions to Brownian motion of molecules, and even laid the foundations, inspired by the work of Bose, for a new type of quantum statistics. It is abstruse to believe that he wished to explain where every atom of a gas is located or where every photon resides in a beam of light. In fact, it was Bose and Einstein who taught us that quantum particles are indistinguishable and it makes no sense to give them different labels when dealing with large numbers of them. What was it, then, that Einstein meant when he said that God does not play dice?

Einstein's deity was nature, the universe, and the laws that govern nature. Events of nature that physics describes were for Einstein occurrences in space-time, only understandable through theories based on space-time. I suspect that there are few people on our planet who have thought about space-time as deeply as Einstein has and consequently very few who can claim they understand "events of nature" as well as Einstein did. I do not claim to be one of them. However, some key points were made entirely clear in the EPR paper and subsequent comments of Einstein.

Einstein demonstrated by his EPR Gedanken experiments that probability laws combined with correlations of events may introduce instantaneous influences at a distance that Einstein's relativity theory had exorcised from physics. He demonstrated connections between quantum superposition and "spooky actions," and his example of the superposition of exploded and unexploded gunpowder has led to a considerable revision of the old quantum mechanics of the Copenhagen school by the current consistent history approach. Only if one adheres to Bell's teachings, that Einstein's elements of reality do not exist, can one still defend Bohr and his interpretation of quantum mechanics. Then, however, one also needs to accept instantaneous influences at a distance. If one believes that the measurement outcomes for entangled photon pairs are entirely random in both Tenerife and La Palma and that no elements of physical reality exist, then instantaneous influences at a distance are the only remedy to explain the correlations. Why is that so? Why can no other physical effect explain the correlations?

The full force of Einstein's logic becomes clear if one realizes that distant correlations cannot be expressed, detected, noticed, or even conceived without the additional concept of a time window, without two distant clocks that are synchronized and used to correlate the events. This fact goes to the heart of the philosophical basis of Einstein's teachings. The Copenhagen school followed Mach and attempted to create a science that dealt exclusively with sense impressions, with data. The EPRB experiments and their interpretations do indeed deal mainly with such data, but Einstein taught us that in addition we need to add the man-made space-time concept. Without that concept, correlations of distant events cannot even be recorded! In a way, correlations are, therefore, at the heart of the space-time concept. This fact shows, according to Einstein, the real difficulty of physics being also a philosophy concerned with the nature of the world. Physics describes reality such as data. In turn, however, we "understand" reality only through its physical description, its description going beyond the mere sense impressions and involving theories.

There is, however, at least one firm ground in this shaky game, as Fine (Fine, 1986) called it. We can carve out alternatives. Correlations can be linked to elements of physical reality, or we have to use instantaneous influences. We can use Einstein's elaborate space-time system to describe correlations in nature, or we can assume that the correlations of different events are a direct expression of natural law that involve instantaneous influences for distant events. Such influences were for Einstein spooky, a matter of the past that did not make any sense, because of his findings about "simultaneity," a concept that is closely related to the word *instantaneous*. EPR experiments show all these complications that go far beyond what we can describe by the rules of ordinary dice games that do not involve the intricacies of space-time. EPR experiments also highlight the problems that the introduction of probability into natural laws brings with it. If some events "happen by chance," they can be only completely correlated to other events that also "happen by chance" through either elements of reality or instantaneous influences. This is how Einstein justified his one liner "Gott wuerfelt nicht." The laws of nature, Einstein's deity, cannot be seen in a complete analogy with dice games. Space-time is special.

John Bell did not think so and pulled the Trojan horse of dice games and instantaneous influences into the court of the physics castle, and it still resides there and unleashes its magic: quantum nonlocalities. The most visible consequences of the faith of many scientists in these nonlocalities are the enormous efforts in the areas of quantum computing, quantum teleportation, and the creation of a variety of entangled qubits by a large and growing number of researchers. The main attraction for all of this research is the possibility of instantaneous influences at a distance that promise massive parallel interactions of qubits and, as a consequence, massive computer power. Will the quantum computer systems that are currently developed work on that basis of quantum nonlocalities, or just according to the well-known principles of conventional information theory? Will quantum computers work in entirely new ways or just the way combinations of digital and analog computers work? I personally have no doubt that elaborate future quantum computer systems will demonstrate a lack of instantaneous influences at a distance and that Einstein was right.

Bibliography

Adenier, G., Khrennikov, A., and Nieuwenhuizen, T. N. (eds.). (2006). *Quantum Theory: Reconsideration of Foundations 3*. AIP Conference Proceedings, Vol. 810. Melville, NY: American Institute of Physics; pp. 437–446.

Aspect, A., Dalibard, J., and Roger, G. (1982). Experimental test of Bell's inequalities using time-varying analyzers. *Phys. Rev. Lett.* 49:1804–1807.

Ballentine, L. E. (1987). Realism and quantum flux tunneling. *Phys. Rev. Lett.* 59:1493–1495.

Bell, J. S. (1964). On the Einstein–Podolsky–Rosen Paradox. *Physics* 1:195–200.

Bell, J. S., Bell, M., Gottfried, K., and Veltman, M. (eds.). (2001). *On the Foundations of Quantum Mechanics*. Singapore: World Scientific.

Bennett, C. H., Brassard, G., Crepeau, C., Jozsa, R., Peres, A., and Wootters, W. K. (1993). Teleporting an unknown quantum state via dual classical and Einstein–Podolsky–Rosen channels. *Phys. Rev. Lett.* 70:1895–1899.

Boole, G. (1862). On the theory of probabilities. *Philos. Trans. R. Soc. London* 152:225–252.

Brown, T. L. (2009). *Bridging Devides*. Urbana: University of Illinois Press.

De Raedt, H. A., Katsnelson, M. I., and Michielsen, K. (2013). Quantum theory as the most robust description of reproducible experiments. Available from arXiv:quant-ph/1303.4574; pp. 1–19.

De Raedt, H. A., and Michielsen, K. (2012). Event-by-event simulation of quantum phenomena. *Ann. Phys. (Berlin)* (Wiley Online Library), pp. 1–18.

De Raedt H. A., Michielsen, K., and Jin, F. (2012). EPRB laboratory experiments: data analysis and simulation. In *Foundations of Probability and Physics 6*. AIP Conference Proceedings, Vol. 1424. Melville, NY: American Institute of Physics; pp. 55–66.

Einstein, A. (1950) *Out of My Later Years*. New York: Philosophical Library; pp. 106–107

Einstein, A. (1954). *Ideas and Opinions*. New York: Three Rivers Press; pp. 18–24.

Einstein, A., Podolsky, B., and Rosen, N. (1935). *Can quantum-mechanical description of physical reality be considered complete? Phys. Rev.* 16:777–780.

Fine, A. (1986). *The Shaky Game*. Chicago: The University of Chicago Press; p. 38.

Gell-Mann, M. (1994). *The Quark and the Jaguar*. New York: W. H. Freeman and Company; p. 170.

Gill, R. D., Weihs, G., Zeilinger, A., and Zukowski, M. (2002). No time loophole in Bell's theorem: the Hess–Philipp model is non-local. *Proc. Natl. Acad. Sci. USA* 99:14632–14635.

Griffiths, R. B. (2002). *Consistent Quantum Theory*. New York: Cambridge University Press.

Hawking, S. (1988). *A Brief History of Time*. New York: Bantam Books.

Hess, K., De Raedt, H. A., and Michielsen, K. (2012). Hidden assumptions in the derivation of the theorem of Bell. *Phys. Scr.* T151:014002 (1–7).

Hess, K., Michielsen, K., and De Raedt, H. A. (2009). Possible experience: from Boole to Bell. *Europhys. Lett.* 87:6007-p1-p6.

Hess, K., and Philipp, W. (2001). A possible loophole in the theorem of Bell. *Proc. Natl. Acad. Sci. USA* 98:14224–14227.

Hess, K., and Philipp, W. (2001). Bell's theorem and the problem of decidability between the views of Einstein and Bohr. *Proc. Natl. Acad. Sci. USA* 98:14228–14233.

Hess, K. and Philipp, W. (2004). Breakdown of Bell's theorem for certain objective local parameter spaces. *Proc. Natl. Acad. Sci. USA* 101:1799–1805.

Hess, K., and Philipp, W. (2006). Der Satz von Bell und das Konsistenzproblem gemeinsamer Wahrscheinlichkeitsverteilungen. *Mathematische Semesterberichte* 53:152–183.

Khrennikov, A. (1997). Non-Archimedean Analysis: Quantum Paradoxes, Dynamical Systems and Biological Models. Dordrecht, the Netherlands: Kluwer Academic Publishers.

Khrennikov, A. (ed.). (2002). *Foundations of Probability and Physics 2*. Växjö, Sweden: Växjö University Press.

Larsson, J. A., and Gill, R. D. (2004). *Bell's inequality and the coincidence-time loophole. Europhys. Lett.* 67:707–713.

Leggett, A. J. (1987). *The Problems of Physics*. New York: Oxford University Press; pp. 165–166.

Leggett, A. J., and Garg, A. (1985). Quantum mechanics versus macroscopic realism: is the flux there when nobody looks? *Phys. Rev. Lett.* 54:857–860.

Mermin, N. D. (2002). Shedding (red and green) light on "time related parameters." Available from arXiv:quantph/0206118v1; pp. 1–5.

Mermin, N. D.(2004). Reply to the comment by K. Hess and W. Philipp on "Inclusion of Time in the Theorem of Bell." *Europhys. Lett.* 67:693–694.

Myrvold, W. C. (2002). A loophole in Bell's theorem? Parameter dependence in the Hess–Philipp model. Available from quant-ph/0205032; pp. 1–18.

Ozawa, M. (2003). Physical content of Heisenberg's Uncertainty relation: limitation and reformulation. *Phys. Lett. A* 318:21–29.

Pascazio, S. (1986). Time and Bell-type inequalities. *Phys. Lett.* 118:47–53.

Pearle, P. M. (1970). Hidden-variable example based on data rejection. *Phys. Rev. D* 2:1418–1425.

Pitovsky, I. (1989). *Quantum Probability: Quantum Logic*, Physica 321. Berlin: Springer Verlag.

Takagi, S. (1997). *Macroscopic Quantum Tunneling*. Cambridge: Cambridge University Press; pp. 171–175.

't Hooft, G. (2006). The mathematical basis for deterministic quantum mechanics. Available from arXiv:quant-ph/0604008; pp. 1–17.

Vorob'ev N. N. (1962). Consistent families of measures and their extensions. *Theory Probab. Appl.* 7:147–163.

Zeilinger, A., et al. (2007). Reported by Minkel, J. R. (2007). Quantum spookiness spans the Canary Islands. *Scientific American*, March 9.

Index